Felix Publishing 2017
www.felixpublishing.com.au
email: info@felixpublishing.com
Print copies available from publisher.

Rocks - Building the Earth
Part of the Series **Adventures in Earth Science**
Other books in the series include:
Exploration Science (Field Geology and Mapping)
Riches from the Earth (Minerals, Mining & Energy)
Changing the Surface (Erosion and Landscapes)
Fossils – Life in the Rocks
A Dangerous Planet (Earth Hazards)
Through Sea and Sky
Beyond Planet Earth (Astronomy)

2017 digital book release
ISBN: 978-0-9946432-1-5
Print Edition ISBN: 978-0-9946432-2-2
Author: Dr P.T.Scott

All illustrations, photographs and videos by the author unless stated. Cover photo: Rocky valley near the Morado Glacier at 3000 metres in theAndes, Chile. Cover design after that of AJS Creative, Brisbane.

Registration:
Thorpe-Bowker +61 3 8517 8342
email: bowkerlink@thorpe.com.au

FELIX
PUBLISHING

ROCKS – BUILDING

THE EARTH

Dr. Peter T. Scott

First released 2017

FELIX
PUBLISHING

To my grandchildren who are
yet to find their own adventures.

About the Author

Dr. Peter Scott is an award-winning teacher of Earth Science of over forty years' experience in both Secondary and Tertiary Education. He holds a Bachelors' Degree, two Masters' Degrees and a Doctorate including many years on his own research in locating and correlating coal measures. He has visited many places of interest including Antarctica, the Andes, the Amazon, North Africa, volcanic islands of the Pacific and Asia, California, northern Europe and remote Australia.

Dr. Scott, in the Andes heading to the Morado Glacier, Chile 2016

Table of Contents

Chapter 1:Rock-Forming Minerals

1.1 Introduction

Minerals are the basic components of the lithosphere or the upper rocky layer of the Earth's crust, and are naturally-occurring, inorganic (i.e. non-living) substances having definite properties and a known chemical composition. They can be formed by crystallisation from molten material (magma or lava) e.g. quartz within igneous rocks, or from hot vapours and solutions e.g. sulfur from volcanic vents, or by evaporation from water solutions e.g. halite (or rock salt). Details of mineral composition and chemistry are given in the companion book RICHES FROM THE EARTH.

Rocks are formed by the combination of minerals by:

- Crystallization from molten material as **magma** deep underground or from **lava** on the surface to form **igneous rocks.**

- Cementation of weathered and eroded minerals or particles containing minerals or crystallization of deposits from mineral solutions as **sedimentary rocks.**

- Recrystallization from the modification and change of existing minerals in rocks due to the application of heat and/or pressure to form **metamorphic rocks.**

As rocks are often identified by their constituent minerals, it is important to be able to identify these rock-forming minerals in both hand specimen and in the field. This is done by identifying minerals by their common properties.

1.2 Minerals and Their Properties

Minerals can be described, classified and identified in terms of their most obvious characteristics or properties:

- **Habit** describes the overall appearance and arrangement of the crystals of the mineral specimen. The most common habits are:

Irregular – crystal faces in many directions
e.g. feldspars

Fibrous – clusters of fine threads e.g. asbestos

Radiating - star pattern of needles e.g. gypsum

Botryoidal - semi-spherical lumps e.g. haematite

Mammillary - similar to botryoidal but with flatter, rounded lumps
e.g. goethite

Pisolitic - pea-like spheres
e.g. bauxite

Foliated – with flat sheets
e.g. muscovite micas

Massive - shapeless with no crystal faces being seen e.g. kaolinite

- **Colour** is shown by the wavelength of light emerging from within the internal structure of the mineral. This may not be reliable in identification, as in many cases a mineral may have different colours under different conditions e.g. quartz could be clear (as rock crystal), white (as vein quartz), smoky grey (as cairngorm), purple (as amethyst), pink (as rose quartz) and other colours depending upon slight impurities of metallic elements.

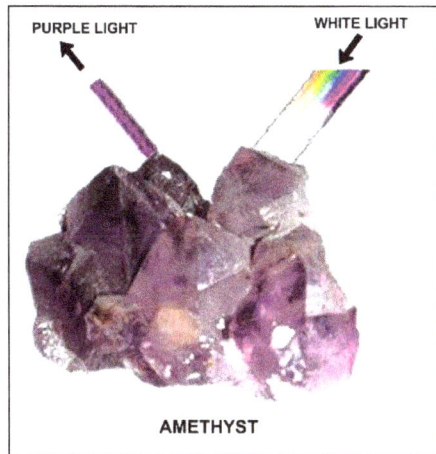

Figure 1.1: Colour as a transmitted wavelength

- **Streak** is the colour of the powdered mineral made by firmly pressing the mineral across a white tile called a streak plate. A black tile is used if the mineral gives a white, cream or light yellow streak. Minerals which are harder than the plate have to be scratched by harder minerals, such as corundum or by a tungsten-tipped scribe or file and then the powder smeared across white or black paper. Streak is usually more reliable than colour e.g. haematite may be red, black and various shades of brown but its streak is always cherry-red.

- **Lustre** is the way which light reflects off the surfaces of the mineral specimen. Lustre can be:

  - Metallic - with a hard, very reflective shine like metals e.g. pyrite (yellow at right), also gold, galena

Figure 1.2: Ore specimen with pyrite and sphalerite

  - Sub-metallic with the reflection similar to metals but not as shiny e.g. sphalerite (black at right).

  - Non-metallic - a great variety of common lustres including:

Adamantine (or brilliant) - sparkling like diamond;
Vitreous - glassy e.g. quartz;
Sub-vitreous - not as glassy e.g. calcite;
Pearly - like a pearl or shirt button e.g. talc;
Silky - shiny & fibrous like silk e.g. gypsum;
Resinous - dull but with a resin-like appearance e.g. white opal; or
Dull - very little reflection; dull like dirt e.g. limonite.

Other terms or combination of terms may be preferred. Some minerals may have several different lustres from one specimen to another because of different habits and a particular specimen may have different lustres on different cleavage planes e.g. feldspars may have a sub-vitreous lustre on one cleavage plane but a dull lustre on the other cleavage plane.

- **Diaphaneity** is the way that light passes through the specimen in normal thickness. This may be described as either:

    Transparent - with light passing through undistorted so that images can be seen through it e.g. topaz

    Translucent - light passes through but is distorted like in bathroom glass e.g. muscovite mica; and

    Opaque - light will not pass through the normal specimen e.g. feldspar.

This property is usually useful only when the mineral is typically transparent or translucent, especially gem minerals such as topaz and chalcedony.

- **Hardness** is the resistance to scratching when tested with a standard set of items having a relative strength. The **Mohs' Scale** was developed by Friedrich Mohs (German: 1773-1839) and uses a set of standard minerals which have been given values of hardness and arranged in order from softest (1) to hardest (10).

When using this scale, test minerals are firmly pressed across a good, flat surface of the unknown specimen to see if they will scratch the surface. For example, if the mineral fluorite will just scratch the unknown mineral, then that mineral will be said to have a hardness of 4. When testing hardness, it is a

MOHS' SCALE of HARDNESS	
HARDNESS	STANDARD MINERAL
1	Soft TALC
2	GYPSUM
3	CALCITE
4	FLUORITE
5	APATITE
6	FELDSPAR
7	QUARTZ
8	TOPAZ
9	CORUNDUM
10	hard DIAMOND

Figure 1.3: Mohs' scale of hardness

good practice to rub the mark left when the test mineral is pressed across the surface - the mark could be traces of powder left behind by a softer mineral. This will rub off, but a true scratch mark will be left behind. If the hardness of a mineral is between two values, then a half value can be given e.g. if the mineral has a hardness between that of fluorite (4) and apatite (5), then its hardness can be given as 4.5.

In the field, a geologist may carry some common items with which to quickly and conveniently test for approximate hardness.

FIELD SCALE of HARDNESS	
HARDNESS	STANDARD MINERAL
2.5	FINGERNAIL
3	COIN
5.5	KNIFE BLADE
6.5	STEEL FILE
7	COMMON QUARTZ

Figure 1.4: A simple hardness scale for field use

Whilst some minerals vary in hardness, others, especially the harder gem minerals, have very specific diagnostic hardness. In the field, a geologist looking for gem materials will pick up a piece of white common vein quartz to use to test hardness on any specimen which has a transparent or translucent diaphaneity i.e. it looks like glass.

- **Cleavage** is the way that some minerals split along natural, flat planes of weakness when gently struck. These planes can be:

  perfect - very smooth and shiny
  good - smooth with some shine
  indistinct (or poor) - flat surfaces with some roughness giving little shine
  no cleavage

As well, cleavage may be described as in:

One direction (basal cleavage) giving small sheets e.g. micas

Two directions giving small steps e.g. feldspars

Three directions giving points e.g. galena

Four directions (octahedral cleavage) giving pyramids e.g. fluorite

Some minerals may grow as well-formed crystals, with faces which look like cleavage planes, but because of their internal crystal structure, these minerals may not have any cleavage at all.

Figure 1.5: Quartz has no cleavage but is often found as crystals with flat faces which may look like cleavage planes

Determining cleavage by inspection is satisfactory for some of the common minerals which have obvious cleavages, such as those with perfect basal cleavages such as micas and topaz, and the three cleavages of galena, pyrites and calcite. It can become confusing. However, as some minerals such as quartz have no cleavage when struck, but do show good flat planes meeting at edges or points due to the original growth of its crystal faces. The only true test for cleavage is to gently strike the specimen and observe the number of non-parallel planes along which it splits.

- **Fracture** is the way the mineral breaks up roughly, but not along smooth planes of weakness. Not a very helpful diagnostic property in general, but there are some useful patterns of fracture which can be a characteristic feature of a particular mineral. These include:

  even - surfaces are generally flat but not as good as cleavage

  uneven -very rough surfaces

  splintery -long, broken parallel fibres

  earthy - rough but with rounded edges like lumps of dirt

  hackly - rough, pointed surfaces in many directions

conchoidal -very sharp edges with concentric circular marks like the edges of broken glass or shell. **Conchoidal fracture** is the most easily identified fracture seen in the field and is typical of quartz, topaz and volcanic glass.

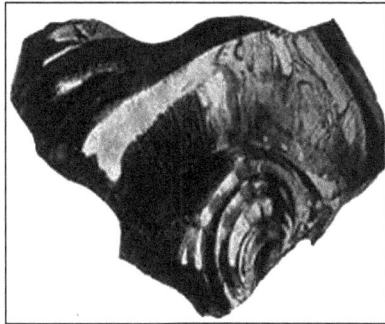

Figure 1.6: Conchoidal fracture in the volcanic glass obsidian (lower right)

- **Crystal family** is a useful diagnostic aid when some specimens have distinct, well-formed crystals, large enough to show distinct crystal faces and the angles between them.

Usually, the shape of the crystal family is defined by the axes (more than one axis), or the framework lines, around which the outer surfaces or crystal faces are built and the angles between these axes. For example, the simplest crystal shape is a cube, which can be defined as being that shape having all three axes equal in length and being at right-angles to each other (i.e. axis a=axis b=axis c all at 90°). These simple shapes may also be further modified.

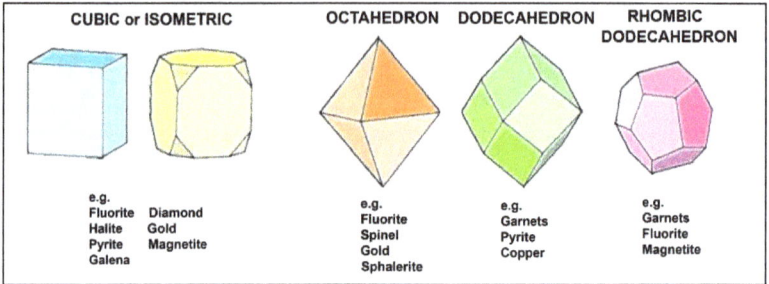

Figure 1.7: The parameters for the cubic crystal family

This identification is only useful if the simple shape of the crystals can be seen and there are no other variations of the crystal form. Complications occur when crystals grow as variations of the simple form or when they are twinned or when they exist as complex aggregates.

Figure 1.8: An aggregate of quartz crystals (x ½)

In very simplified terms, and for ease of observational identification, the crystal families are summarised in the following table:

NAME	DESCRIPTION	LATTICE (Framework)	EXAMPLES
Cubic Family	Axes are equal length and are all at 90° to each other	$a = b = c$ all angles = 90°	Halite (sodium chloride) galena (lead sulfide) pyrite (iron sulfide) fluorite (calcium fluoride) diamond (the element carbon) magnetite iron oxide)
Tetragonal Family	Two axes are equal in length and the other shorter or longer, with all axes at 90 degrees	$a1 = a2 \neq c$ all angles = 90° ($\neq$ means "not equal to")	Zircon (zirconium silicate) Rutile (titanium dioxide) cassiterite (tin oxide) chalcopyrite (copper iron sulphide)
Hexagonal Family	Two horizontal axes equal in length and at 120° to each other and an unequal vertical axis at 90° to these two.	$a1 = a2 = a3 \neq c$ angle between a axes & c = 90° angles between a axes all = 60°	beryl (beryllium aluminium silicate) graphite (carbon) apatite (calcium fluoro-phosphate) Chile saltpetre (sodium nitrate) tourmaline (complex silicate with boron, aluminium, alkali metals iron and magnesium)

Orthorhombic Family	All axes are unequal in length but are all at 90° to each other	$a \neq b \neq c$ all angles = 90°	aragonite (calcium carbonate) barite (barium sulfate) olivine (iron-magnesium silicate)
Monoclinic Family	All axes are unequal in length with one axis vertical, another at 90° to the vertical and the third axis at an oblique angle to the plane of the other two axes	$a \neq b \neq c$ angle between a & b and b & c = 90° angle between a & c > 90°	Muscovite (complex aluminosilicate) biotite (complex aluminosilicate) orthoclase (potassium aluminosilicate) gypsum (calcium sulfate with water) hornblende (complex aluminosilicate) augite (complex aluminosilicate) malachite (basic copper carbonate)
Triclinic Family	All axes are unequal and none of the angles are at 90°.	$a \neq b \neq c$ all angles ≠ 90°	plagioclase (calcium to sodium alumino-silicate) kaolinite (aluminium silicate), rhodonite (manganese silicate)

Table 1.1: The main crystal families

Some 3-dimensional classifications use crystal system rather than crystal families. This is a more technical classification and involves a more complex identification, involving planes, or mirrors of symmetry and amounts of rotation around each axis as rotational symmetry. Consequently, there is often considerable confusion when discussing types of crystal shapes. For example, there are six crystal families and seven crystal systems, the latter including an additional member, the trigonal system.

The **trigonal crystal system** with all axes equal and all angles equal (but not 90⁰). Some members show external hexagonal shape (e.g. quartz) and others a rhombic shape (e.g. calcite). It is a subset of the Hexagonal System. Examples of trigonal minerals include:

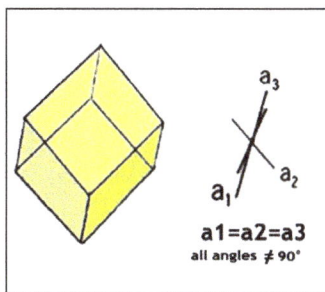

a1=a2=a3
all angles ≠ 90°

> quartz (silicon dioxide)
> calcite (calcium carbonate)
> dolomite (calcium magnesium carbonate)
> corundum (aluminium oxide)

The term **rhombohedral system** is often used synonymously for the trigonal system, but the rhombohedral system is defined by its crystal lattice rather than outward appearance and is a subset of the trigonal lattice system. It has all sides equal including the c-axis, unlike the hexagonal system where the c-axis is a different length and the angle between the c-axis and the others is not 90^0. Calcite and dolomite can sometimes have this shape. A summary of the main terminology used in crystallography is given below:

CRYSTAL FAMILY	CRYSTAL SYSTEM	LATTICE SYSTEM
Triclinic		Triclinic
Monoclinic		Monoclinic
Orthorhombic		Orthorhombic
Tetragonal		Tetragonal
Hexagonal	Trigonal	Rhombohedral
		Hexagonal
	Hexagonal	
Cubic		Cubic

Table 1.2: Different systems of crystallography

- **Specific gravity (S.G.)** is the density of the mineral compared to that of water which is 1.0 gram per cubic centimetre.

i.e. Specific Gravity = $\dfrac{\text{Density of mineral}}{\text{Density of Water}}$

The density of the mineral is calculated by dividing its mass usually in grams, by its volume in cubic centimetres (note: 1 c.c. approximately equals 1 millilitre). Alternatively, this can be found by dividing the mass of the mineral weighed in air by its apparent loss in mass when completely submerged in water using Archimedes' Principle. Since it is a ratio, specific gravity, has no units of measurement e.g. the S.G. of gold is 19.3. **Heft** is a rough comparison of the heaviness of a mineral which may be useful for quick identification of some minerals. Heft may be ranked as heavy, medium or light e.g. calcite has medium heft, but barite, which it closely resembles, has a heavy heft.

- **Chemistry** refers to the chemical composition of the mineral i.e. what elements and chemical bonding make up the mineral.

Some simple chemical tests can be applied in the field to identify some common chemical families e.g. acid on carbonates such as calcite, dolomite and magnesite, will give bubbles of odourless carbon dioxide gas. Strong acid on sulphides, such as pyrite and galena, will give a bad odour of rotten eggs due to hydrogen sulfide gas and this test can be used to distinguish fool's gold (pyrite) from real gold which has no reaction with the acid.

- **Specific properties** refer to any unique features of the mineral which may be useful in quick identification. e.g.

    - magnetite can be attracted to a magnet

    - talc feels slippery

    - fluorite, autunite and some other minerals are fluorescent

    - micas are flexible and elastic

19

- pitchblende, uraninite and monazite are radioactive

- kaolinite has an earthy taste

- halite (rock salt) tastes salty

- tourmaline becomes charged electrically when heated

- Iceland spar is a clear form of calcite which gives double refraction i.e. splits light up into two beams

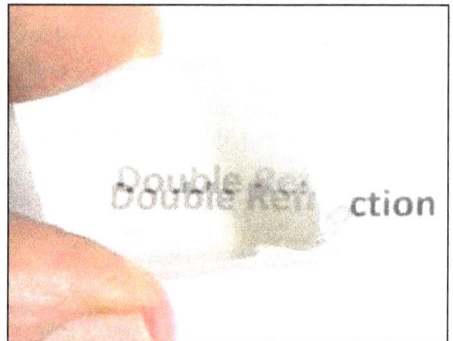

Figure 1.9: Iceland spar showing double images by double refraction

In the field, looking for specific minerals with which to identify rocks, the geologist usually uses a range of common properties such as colour, hardness and habit but in the field most minerals and their rocks are often badly changed by the weather so the task becomes harder.

1.3 Some Common Minerals

There are many other chemical compounds which form

minerals. These, along with the silicates form the great majority of rocks (which are mixtures of minerals), ores (economically valuable minerals), gemstones (valuable metals and minerals used for adornment) and other earth materials. The most common groups of minerals based upon their chemistry are given in the next table:

SILICATES silicon, oxygen & other elements	OXIDES oxygen & other elements	CARBONATES carbon, oxygen & other elements	PHOSPHATES phosphorus oxygen & other elements	SULFATES sulfur oxygen & other elements	SULFIDES sulfur & other elements	HALIDES chlorine, bromine, iodine & other elements
Olivines $MgSO_4$ to $FeSO_4$	Quartz SiO_2	Calcite $CaCO_3$	Apatites e.g. $Ca_5(PO_4)_3$ (F,Cl,OH)	Gypsum $CaSO_4.2H_2O$	Pyrite FeS^2	Halite $NaCl$
Pyroxines e.g.Augite (Ca,Na) (Mg,Fe,Al,Ti) $(Si,Al)_2O_6$	Haematite Fe_2O_3	Dolomite $CaMg(CO_3)_2$	Monazite $(Ce,La,Y,T)PO_4$	Barite (Baryte) $BaSO_4$	Sphalerite ZnS	Fluorite CaF_2
Amphiboles e.g. Hornblende $Ca_2(Mg, Fe, Al)_5 (Al, Si)_8O_{22}(OH)_2$	Magnetite Fe_3O_4	Magnesite $MgCO_3$	Autunite $Ca(UO_2)_2$ $(PO_4)_2\cdot$ $10\text{-}12H_2O$	Anhydrite $CaSO_4$	Galena PbS	Cryolite Na_3AlF
Micas e.g. Muscovite $KAl_2(AlSi_3O_{10})$ $(F,OH)_2$ Biotite $K(Mg,Fe)_3AlSi_3O_{10}(OH)_2$	Limonite $FeO(OH)\cdot$ nH_2O Where n = any number	Siderite $FeCO_3$	Turquoise $CuAl_6(PO_4)_4$ $(OH)_8\cdot5H_2O$	Epsomite $MgSO_4\cdot7H_2O$	Chalcopyrite $CuFeS_2$	Bromargyrite $AgBr$
Plagioclases $NaAlSi_3O_8$ - $CaAl_2Si_2O_8$	Rutile TiO_2	Cerussite $PbCO_3$				
Orthoclase $KAlSi_3O_8$	Corundum Al_2O_3					
Quartz (also an oxide) SiO_2	Ice H_2O					

Table 1.3: Some of the common chemical groups of minerals

On the following pages are some of the most common minerals found in the field. A good knowledge of these minerals and their characteristic properties is useful in identifying them and the rocks or deposits in which they may be found. It is important to remember that these minerals will alter through reaction with natural gases and solutions (weathering) after they are exposed, so identification is often more difficult unless fresh specimens are found.

Muscovite Mica	Biotite Mica	Orthoclase Feldspar

Silver basal cleavage	Black, basal cleavage	Pink, hardness 6 2 cleavages

Plagioclase Feldspar	Gypsum	Quartz

Cream, hardness 6, 2 cleavages	White, waxy, hardness 2	Glassy, hardness 7,

Olivine

Olive, glassy, hardness 7

Augite

Green-black, stubby 2 cleavages

Hornblende

Black, hardness 6
2 cleavages

Kaolin

White, dull, soft

Fluorite

Green, glassy, hardness 4

Halite

Clear, glassy, tastes salty

Calcite

Hardness 3 3 cleavages

Barytes

Heavy, 3 cleavages

Azurite (Blue) & Malachite (Green)

Glassy to dull, hardness 3-4

Pyrite	Galena	Sphalerite

Metallic yellow, 3 cleavages hardness 6-6.5	Heavy, metallic, 3 cleavages hardness 2.5	Grey, shiny, sub-metallic hardness 4

Some of the major diagnostic properties of the most common minerals are given in the following table:

MINERALS	PROPERTIES									
	FORM	CLEA	FRACT	HARD	COL.	LUST	STREAK	S.G. & CHEM.	CRYSTAL SYSTEM	OTHER
QUARTZ	crystal	None	Conch.	7	varies	glassy	white	2.65 SiO_2	Hex.	gems
ORTHOCLASE	blocks	2 @ 90^0	uneven	6	pink	glassy	white	2.57 $KAlSi_3O_8$	Mono.	simple twins
PLAGIOCLASE	blocks	2 @ almost 90^0	uneven	6-6.5	white	glassy	white	2.62 $NaAlSi_2O_8$ to $CaAl_2Si_2O_8$	Tri.	multi twins
MUSCOVITE MICA	Sheets	basal (1)	uneven	2-2.5	silver	glassy	white	2.76-3.00 Silicate of Al, OH & F	Mono.	bends
BIOTITE MICA	sheets	basal (1)	uneven	2.5-3	black	glassy	white-brown	2.80-3.10 Silicate of K, Fe, Mg, Al, OH &F	Mono	bends
HORNBLENDE	Blocks	2 @ 56^0	uneven	5-6	black	glassy	green-brown	2.90-3.40 Silicate of Fe, Mg, Na, Al & OH	Mono	long laths
AUGITE	crystal	2 @ 90^0	uneven	5-6	black-green	glassy	grey green	3.20-3.50 Silicate of Ca, Mg, fe, al	Mono	stubby
OLIVINE	grains	none	Conch.	6.5-7	olive green	glassy	white	3.20-4.40 $(FeMg)$-SiO_4	Ortho.	gels with acid
CALCITE	blocks	3	even	3	white	glassy	white	2.70 $CaCO_3$	Hex.	co_2 with acid
BARITES (BARYTE)	blocks	3	even	3-3.5	white	glassy	white	4.30-5.00 $BaSO4$	Ortho.	heavy
KAOLINITE	Massive	basal	uneven	2-2.5	white	dull	white	2.6 $Al_2Si_2O_5$.$(OH)_4$	Tri.	earthy taste
HALITE ("salt")	blocks	3 cubic	even	2.0-2.5	white	glassy	white	2.17 $NaCl$	Cubic	salty taste
HAEMATITE	varies	none	uneven	5.5-6.5	red	dull - metal	red	5.26 Fe_2O_3	Trig.	red soils
MAGNETITE	blocks	3 cubic	uneven	5.5-6.6	grey-black	metal	black	5.17-5.18 Fe_3O_4	Octahed	magnet.
LIMONITE	varies	none	uneven	4.0-5.5	yellow	dull	yellow	2.9-4.4 $FeO(OH)$·nH_2O	Amorph.	Yellow pigment
PYRITE	blocks	3	uneven	6.0-6.5	gold	metal	green-black	4.95-5.10 FeS_2	Isomet.	fool's gold
CHALCOPYRITE	blocks	indistinct	uneven	3.5	yellow	metal	Green-black	4.1-4.3 $CuFeS_2$	Tetrag.	magnet. with heat
GALENA	blocks	3 cubic	Sub-chonch.	2.50-2.75	lead grey	metal	lead grey	7.2-7.6 PbS	Cubic	heavy
SPHALERITE	blocks	1	uneven	3.5-4.0	brown-black	Sub-metal	brown	3.9-4.0 $ZnFeS$	Iso.	Shows fluores.

Table 1.4: Properties of some of the common rock-forming minerals

Chapter 2: Igneous Rocks - The Beginning

2.1 Introduction

Igneous rocks were the first rocks to form on the Earth, coming from the solidification of molten material when the young Earth cooled. Today, igneous rocks are still being formed on the surface from the lava of active volcanoes and deep below the surface within the Earth's crust from upwelling molten material known as **magma** (from the Ancient Greek meaning thick, sticky substance).

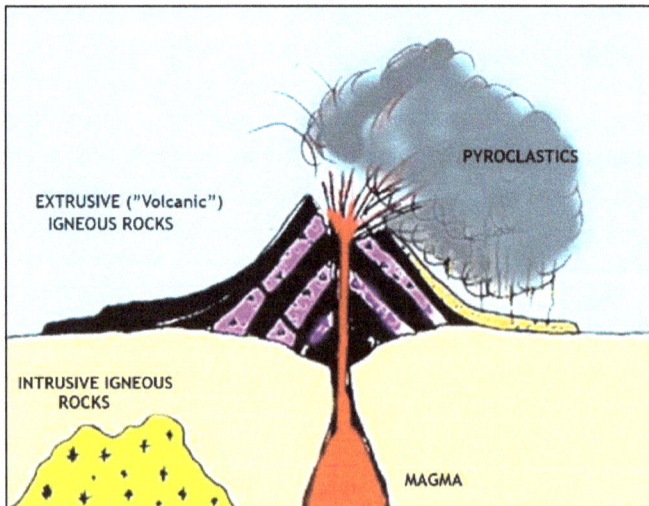

Figure 2.1: A simplified view of major igneous rock types

Magma cools and crystallises deep below the surface forming intrusive igneous rocks which usually have well-formed crystals. When magma comes to the surface it cools quickly, forming the finely crystalline extrusive igneous rocks typical of volcanic areas. If the extrusion is associated with release of great pressure, explosive **pyroclastic** rocks would be formed.

Light-coloured igneous rocks are also called **felsic** igneous rocks because they have much feldspar and quartz or silica. Dark-coloured igneous rocks are also called **mafic** igneous rocks because they contain many minerals containing magnesium and iron (+ferric) such as hornblende, augite and biotite.

Pyroclastic eruptions produce igneous rocks having broken and angular particles of different sizes. All pyroclastic particles, irrespective of size and shape are called **tephra**:

TEPHRA		SIZE	RESULTING ROCK
Blocks		> 25.6 cm	volcanic agglomerate
Bombs		3.2 to 25.6 cm	volcanic agglomerate
Lapilli		4.0mm to 3.2 cm	volcanic breccia
Ash		0.062 mm to 4.0mm	tuff
Dust		<0.062 mm	tuff

Table 2.1: Sizes of pyroclastic material ejected during eruptions

WIDTH = 30 cm.

Figure 2.2: A volcanic bomb - note its aerodynamic shape (Photo: USGS)

2.2 Intrusive Igneous Structures

Magma, from very deep sources may extrude up into crustal rock as a semi-molten or molten mass which may cool deep underground to form several types of structures. These are called **intrusive igneous** structures or plutons, from Pluto the Roman god of the underworld. When they cool, they may be brought to the surface due to uplift and erosion, and their size of exposure often determines their name. Some of the most common are:

- **Batholiths** have an area of exposure greater than 60 square kilometres.

- **Stocks** have an area of less than 60 square kilometres and circular in shape.

- **Bosses** are very small, circular in shape and with vertical sides.

Often the magma may penetrate higher into the crust to cool at shallow depths, where the fluid hot crystalline mush can flow along planes of weakness such as near-vertical cracks and horizontal bedding planes. Upon cooling, these form smaller structures such as:

- **Dykes** are intrusive structures which cut across, or are discordant to older rock layers, and are often near vertical. They are thin, varying from a few centimetres to several metres and look like a wall when exposed. Conversely, they may weather to clays which are easily eroded out from the hardened surrounding rocks so a dyke may be seen as a long, narrow cutting.

- **Sills** are intrusive structures which form within older layers of rock, and are often almost horizontal to existing rock layers i.e. they are concordant. They are often tabular and are seen as a long, narrow slab within rock strata.

- **Laccoliths** are igneous structures which intrude between the layers of the overlying rock and push them up into domes.

- **Lopoliths** are igneous structures which intrude between layers of the overlying rock and then sink in the middle giving a basin-like appearance.

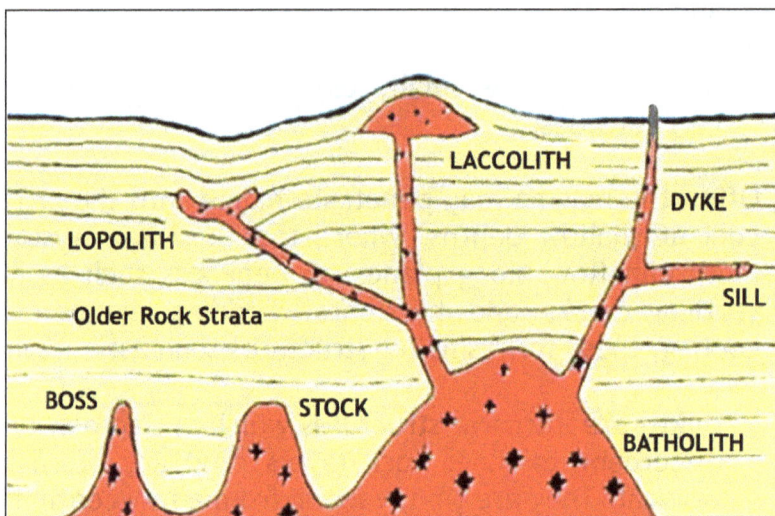

Figure 2.3: common intrusive igneous structures

In the field, deep intrusive igneous rocks can be identified by having large, interlocked crystals, shallow depth Intrusive igneous rocks have some large crystals called **phenocrysts** embedded in a background or groundmass or **matrix** of smaller crystals.

Figure 2.4: Mt. Tibrogargan, one of the Glasshouse Mountains, Queensland, Australa. It is a volcanic plug of trachyte-rhyolite

Figure 2.5: The remains of a small, eroded granite batholith, almost in cross-section near Moruya NSW, Australia.

Figure 2.6: The Breadknife, an eroded dyke in the Warrumbungle Mountains, NSW, Australia.

Figure 2.7: A dolerite sill (outlined) in Cradle Mountain, Tasmania, Australia

In the field, these structures may be very badly weathered and lie under a flat surface or they may be a prominent feature standing out from the general relief. Often there are secondary minerals nearby in the local or country rock, or in long twisting veins of minerals, such as quartz, gold, silver, galena, pyrite, gemstones. Local stream gravels may also contain some of the more resistant minerals such as gold and some gemstones.

2.3 Extrusive Igneous Structures

Volcanic activity at the Earth's surface can produce many different types of structures and rock compositions. Lava which comes from magma with very little quartz content but considerable amounts of mafic minerals, called a **basic magma**, will often rapidly flow across the surface, over plains and through valleys and

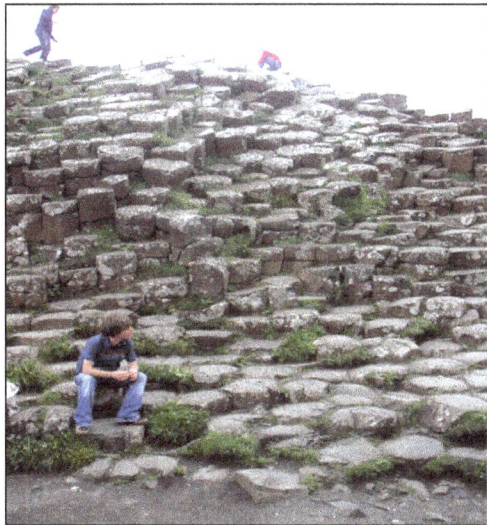

Figure 2.8 Hexagonal basalt columns of the Giant's Causeway, Northern Ireland

then cool as a lava flow. Lava may also form as a hard cap on softer rock, giving then a flat top or plateau. Lava can come from long fissures or cracks as fissure basalts, or it may erupt from a single volcanic vent. When lava cools, it may do so around cooling centres, contracting inwards to form columnar jointing.When it comes from a volcanic eruption, the lava may build up as many layers and form a large **shield volcano**.

Rhyolitic dome volcanoes come from magma rich in silica (quartz), are more explosive and their eruptions can be sudden and destructive. These volcanoes can be very deceptive and often look like non-volcanic, irregularly-shaped hills, but can explode without warning.

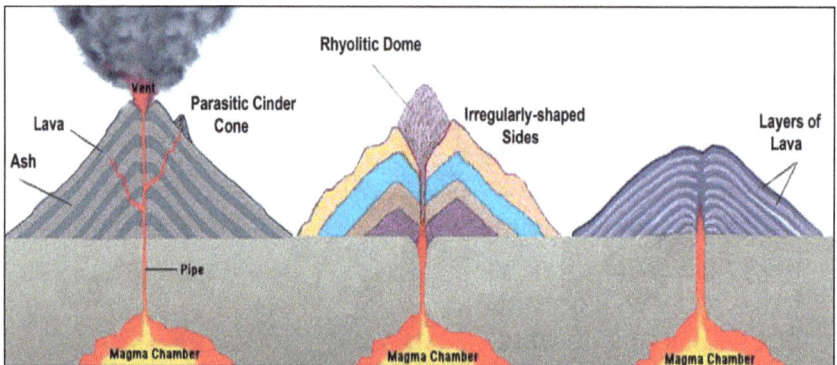

Figure 2.9: The three major types of volcanoes. At left is the stratovolcano with uniform slopes of ash and lava e.g. Mt. Fuji (Japan) and Mt. Taranaki (New Zealand). In the centre is the explosive rhyolitic volcano with irregular sides and often a steep, central spire pushed out of the volcano's vent e.g. Mt. Tarawera (New Zealand). At right is the shield volcano made from successive layers of fast-flowing mafic lava e.g. Mauna Loa (Hawaii).

The most common type of volcano is the **stratovolcano** (or composite volcano) which has a well-known conical shape and is made up from successive layers of both ash and lava. Initially they may explode and deposit extensive layers of ash followed by an eruption of slow-flowing, blocky lava. This cycle is repeated and the volcanoes grow in size producing the uniform slopes of the classic volcanic cone.

2.4 Magma Formation

It was once thought that magma originated within the Earth's crust, which extends to a depth of about 35km, when conditions of temperature and pressure were suitable for melting were reached. Lately, evidence suggests that magma does not come from complete re-melting of crustal rocks, but originates from below the crust in the mantle down to depths of several hundred kilometres.

The theory of plate **tectonics** is useful in explaining the formation of magma and of many igneous events. In this theory, the surface of the Earth is considered to be covered with several large plates, each moving slowly with respect to the others. The plates are about 50 - 150 km thick and thus contain the crust and part of the upper mantle, called the lithosphere. These plates are produced by deep mantle magma coming up at ocean ridges and then moving outwards across the Earth's surface, eventually sinking at an ocean trench. The system can be considered as a giant conveyor belt moving across the surface of the globe carrying the continents with it.

As a plate sinks, or subducts at a trench, ocean sediment, crustal rocks and **ultra-basic** mantle material, which is very rich in iron-magnesium minerals but has little quartz, partially melts. This material, theoretically called **pyrolite** rises through the mantle and separates into different mineral fractions at various depths, producing different magmas:

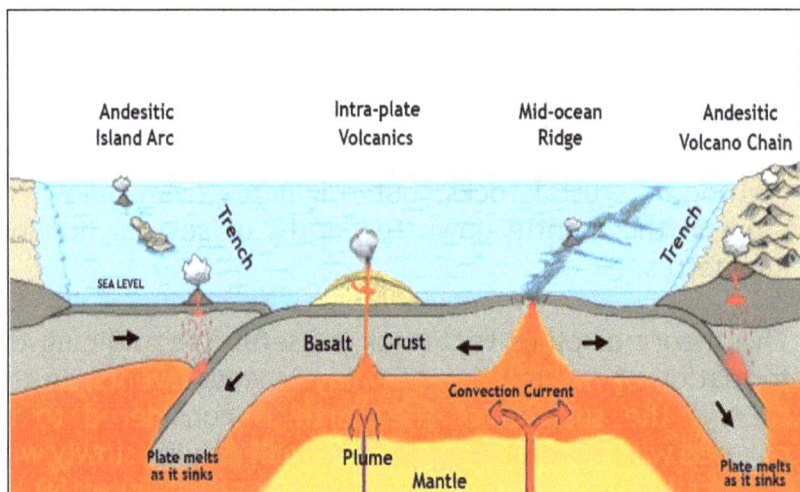

Figure 2.10: A cross-section through the upper part of the Earth across showing some of the volcanism produced at various locations

- Basaltic magma may rise up through weaknesses in the crust and produce vast flood basalt plains. These can be found in the Deccan of India, Siberia, Greenland, the southern tip of South America, Northern Ireland and other places. It may also come up from magma due to melting over a mantle plume within a plate, such as at Hawai'i. Here, a special type of basaltic magma, called **hawaiite** forms the great shield volcanoes which make up the islands of Hawai'i.

- Andesitic magma is produced at subduction zones where the downward movement of the plate also carries sediments from the oceanic trench as well as parts of the basaltic crust. The trench sediments provide more quartz and water for the mix when the melting temperature is reached. This means that andesitic magma has more quartz and volatile components which makes the volcanoes formed as the magma reaches the surface, more explosive. **Andesite** is the typical rock from the lava of these volcanoes;

- Rhyolitic magmas develop beneath continental plates and have a complex composition, because of the variety of rock materials within the continental plate. There is much more quartz in the magma, so when it reaches the surface, the resulting volcanism is often very explosive giving lavas of **rhyolite** and a considerable amount of tephra.

2.5 Composition of Igneous Rocks

Igneous rocks contain minerals which crystallise from magma under different conditions to form a great variety of rock types. The main controls are those of:

- temperature as minerals will have different melting points or specific temperatures at which they will form

- pressure which also affects the crystallization point as well as the size of the crystals

- chemical environment which is the source of the ions of the chemical elements or groups available

This process often is complex, because minerals often do not crystallise all at once. Some crystals interact with the residual liquid which has been left over after the first minerals have solidified, to form new crystals at a lower temperature.

Certain minerals form at particular temperatures called their crystallisation point. Further reaction with the residual liquid will determine the rest of the rock's composition. Continuous Reactions can occur when ions within the residual liquid move into existing crystals and replace ions already within the crystal lattice. For example, in plagioclases, more sodium ion replaces the calcium ions of the crystal as the magma's temperature drops. Discontinuous reactions are more complex, and involve the complete or partial dissolving of previously-formed crystals into the residual liquid, and then their recrystallisation as new minerals at a lower temperature. Thus, olivine may dissolve, react with the surrounding mixture of ions and crystallise as a pyroxene.

As the magma cools, a wide range or series of minerals can solidify and form igneous rocks of varying composition. **Bowen's Reaction Series** is named after **Norman Bowen** (Canadian: 1887-1956), and can be used to explain why certain minerals are found within certain igneous rocks and so used as indicators of temperature. It also explains why some minerals are found together in some rocks and others are not.

From this chart, note that the:

- Minerals of **discontinuous reactions** are rich in iron and magnesium (i.e. are mafic silicates). They are usually dark green or black in colour.

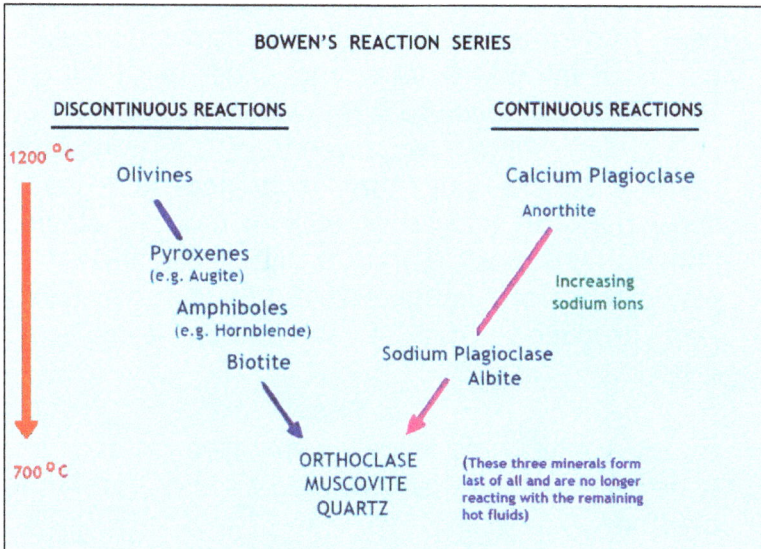

BOWEN'S REACTION SERIES

DISCONTINUOUS REACTIONS   CONTINUOUS REACTIONS

1200 °C
    Olivines   Calcium Plagioclase
                                    Anorthite

    Pyroxenes
    (e.g. Augite)   Increasing
    Amphiboles   sodium ions
    (e.g. Hornblende)
      Biotite   Sodium Plagioclase
                     Albite

                ORTHOCLASE   (These three minerals form
700 °C   MUSCOVITE   last of all and are no longer
                QUARTZ   reacting with the remaining
                                   hot fluids)

Figure 2.11: Bowen's Reaction Series for minerals in igneous rocks

- Minerals of **continuous reactions**, and those at the final stage (orthoclase, muscovite, quartz) are light in colour (i.e. are felsic silicates).

- Minerals which are horizontally opposite (e.g. olivine and anorthite) and which are close together (e.g. augite and olivine) are compatible, and are often formed together within the same rock. For example, at about 1000°C, a basic magma (low in quartz) would cool quickly, preventing further reaction, to produce basalts

39

- rich in olivine, augite and calcium plagioclase (anorthite). At a lower temperature, an acidic magma (rich in quartz) would produce granites, having large amounts of free quartz, muscovite, and orthoclase with some biotite and sodium plagioclase (albite). Some igneous rocks may look identical, but they may have been formed under different conditions. For example, basalt (a mid-plate lava) and andesite (a subduction zone lava) can look the same, and both can be found as lava flows, but an examination of the mineralogy of these two rocks under the petrological microscope will show that the mid-range sodium/calcium plagioclase mineral (andesine) is present only in the andesite. This would therefore indicate that the andesite lava flow was produced from a subduction zone volcanic eruption.

- Minerals which are widely separated on the vertical scale are usually incompatible and rarely found together.

The effects described by Bowen's Reaction Series can be best seen if the rock is sliced, then ground thinly to transparency and cemented to a glass slide which is then placed under a petrological microscope and viewed in polarised light between two polarising filters.

In the left hand slide, showing a discontinuous reaction, notice how the amphibole has reacted and reformed as the biotite and the pyroxene have reacted and reformed as an amphibole. The edges of these growths are not uniform, showing how the first mineral has re-dissolved and then formed a new mineral on the outside. On the right, the reaction has been continuous, i.e. as the crystals formed, one plagioclase (e.g. a calcium variety) has been regularly replaced by another (e.g. a sodium variety).

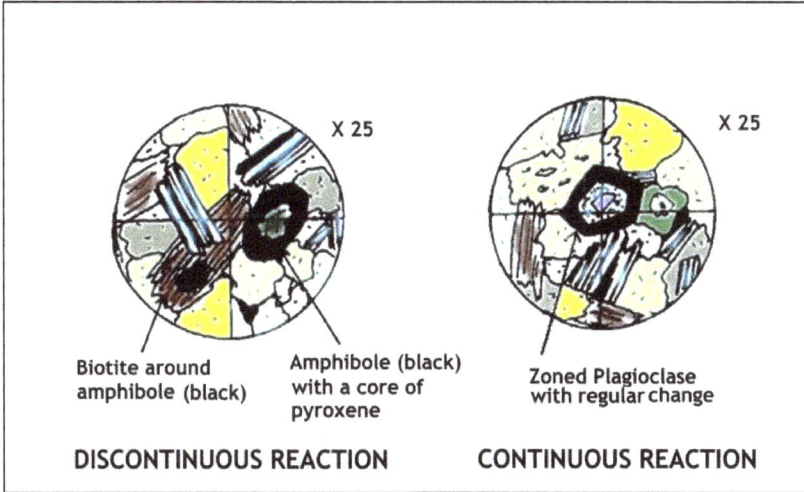

Figure 2.12: Diagram showing some of the different reactions in igneous rocks

2.6 Mineral Textures

Igneous rocks may be described by their:

1. Environment of formation, giving the location of where the rock has cooled from the magma. In this concept, igneous rocks can be either:

   - **extrusive** (volcanic) which are cooled quickly on the surface from molten lava e.g. basalt

   - **intrusive** which are formed below or deep below the surface from the cooling of magma e.g. granite

2. Texture is the geometrical arrangement and shape of the crystals within the rock. Crystal size also depends upon depth and rate of cooling. Large, well-shaped crystals form when slowly cooled at great depth, small crystals form when cooled quickly, and when cooling is rapid, no crystals form, giving a volcanic glass.

When describing a rock's texture, the main types are:

- **Phaneritic** in which all crystals of minerals are large, well-formed and are usually interlocked e.g. as in granite.

Figure 2.13: Granite in hand specimen (left) and in thin section showing interlocked crystals (cross polars X 50)

- **Porphyritic** with large, well-formed crystals called **phenocrysts** set into a background **matrix** or groundmass, of finer crystals. Regardless of the size of the phenocrysts, all porphyries have phenocrysts which are very distinctly larger than the crystals within the groundmass. The difference in sizes is due to the difference in cooling rates - the phenocrysts formed first and slowly crystallised, giving larger crystals as well as very good shape. Next the crystals of the groundmass cooled quickly. Porphyries are intrusive igneous rocks which usually form relatively close to the surface. Porphyries are usually named after the main phenocryst mineral e.g. hornblende porphyry.

Figure 2.14: Hornblende porphyry in hand specimen (left) and in thin-section showing a large phenocryst of hornblende (x 25 cross polars)

- **Aphanitic** with no crystals generally visible to the naked eye. Under the microscope, and with **polarized light**, well-formed crystals may be seen e.g. basalt. In thin-section, the long plagioclase crystals are often aligned, which indicates the direction of the lava flow in the line of the laths, when the basalt cooled.

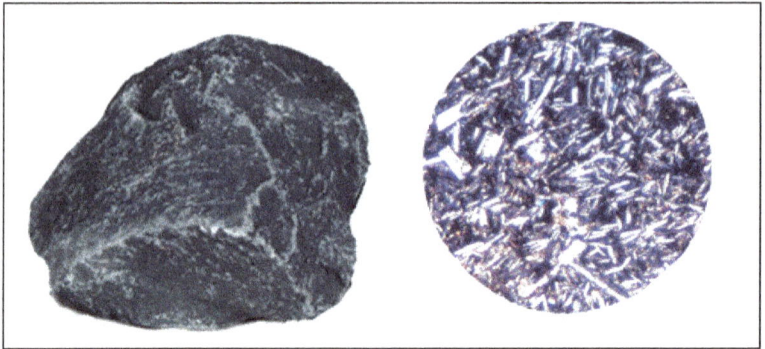

- **Holohyaline** (or glassy) is when the lava cooled so quickly e.g. flowing into water, that crystal did not form at all e.g. obsidian and other volcanic glasses.

Figure 2.16: Obsidian, a volcanic glass in hand specimen (left) and in thin section (x 25 cross polars)

3. Mineral composition if crystals can be clearly seen, then it can be identified by its constituent minerals. In most cases, the main or **primary minerals** are used for identification. These consist of the most common minerals which make up the bulk of the rock (e.g. olivines, amphiboles, pyroxenes, feldspars, micas and quartz). **Accessory minerals**, which are formed at the same time but occur in small amounts (e.g. magnetite, chromite, apatite and sphene) are also seen. Other minerals which may be found within the rock are the **secondary minerals**. These have been formed later by chemical weathering or alteration of the existing minerals. They can include clays, chlorite, calcite, haematite and serpentine.

Based on an idea from chemistry that non-metal oxides (e.g. carbon dioxide) will react with water to give acids (e.g. carbon dioxide + water = carbonic acid) and that metal oxides (e.g. calcium oxide) react with water to form bases (e.g. calcium hydroxide + water = calcium hydroxide), igneous rocks can also be classified by the amount of quartz (silicon dioxide – a non-metal oxide) present:

- **Acid igneous rocks** have high quartz content, usually above 63% quartz. The name was derived from the concept that oxides such as silica (SiO_2) would dissolve in water to produce an acid.

- **Intermediate igneous rocks** will have only a middle range of quartz, usually between 63% and 52%.

- **Basic igneous rocks** have little quartz, perhaps about 45% to 53% silica.

- **Ultra-Basic Igneous Rocks** have less than 45% silica.

Acid igneous rocks are also light in colour because of the high amounts of quartz and feldspars present (i.e. they are felsic) and basic igneous rocks are darker in colour having many iron and magnesium (mafic) minerals. For example, the acid intrusive igneous rock granite may have a composition of:

45%	orthoclase
25%	quartz
15%	plagioclase
10%	amphibole (e.g. hornblende)
and 5%	biotite.

In addition, there may be some minor muscovite and accessory minerals such as magnetite.

The Table 2.1 below gives an approximation of some of the most common igneous rocks. In it one can determine the composition of (say) basalt by looking at the percentage (by area) of the minerals in the vertical column above the term "basalt & basaltic glasses" an estimate would be approximately:

55%	pyroxene (e.g. augite)
25%	plagioclase feldspar
10%	olivine
5%	amphibole (e.g. hornblende)

and less than 5% biotite.

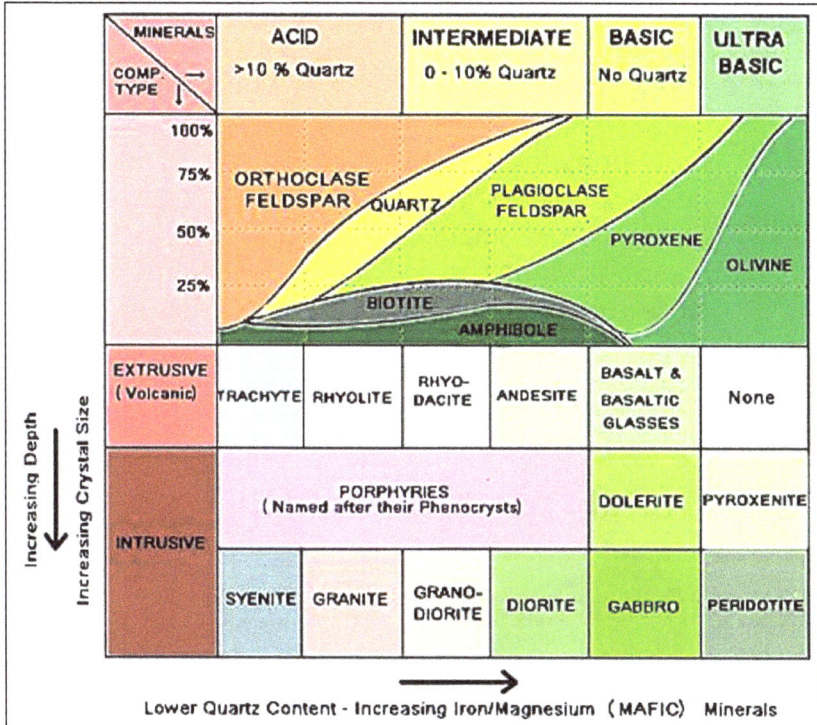

MINERALS COMP. TYPE →	ACID >10 % Quartz		INTERMEDIATE 0 - 10% Quartz		BASIC No Quartz	ULTRA BASIC
100% 75% 50% 25%	ORTHOCLASE FELDSPAR	QUARTZ	PLAGIOCLASE FELDSPAR		PYROXENE OLIVINE	
		BIOTITE AMPHIBOLE				
EXTRUSIVE (Volcanic)	TRACHYTE	RHYOLITE	RHYO-DACITE	ANDESITE	BASALT & BASALTIC GLASSES	None
INTRUSIVE	PORPHYRIES (Named after their Phenocrysts)				DOLERITE	PYROXENITE
	SYENITE	GRANITE	GRANO-DIORITE	DIORITE	GABBRO	PERIDOTITE

Increasing Depth

Increasing Crystal Size

Lower Quartz Content - Increasing Iron/Magnesium (MAFIC) Minerals

Table 2.1: Simplified composition of some common igneous rock

When the minerals cannot be seen with the naked eye, or under the petrological microscope, the rock may be finely ground down to a powder, which is then strongly heated to oxidise all of the minerals into metal oxides. Then the mixture is analysed chemically by testing for each element so that the relative percentages are found and then these results are related to the known compositions of the suspected minerals.

2.7 Internal Rock Structures

In addition to mineral crystals, extrusive igneous rocks may also show some internal features due how they were formed during the volcanic eruption. These may include:

- **Flow lines** in rocks are formed from lavas show as linear or curved markings due to concentrations of material or small aligned crystals within the composition of the rock. These lines are generally close together and follow the direction of the lava flow.

Figure 2.17: Rhyolite in hand specimen (left) showing flow lines and in thin section (x 25 cross polars)

- **Vesicles** are round to elongated gas bubbles formed within the hardening lava of the lava flow. Elongated vesicles are due to the dragging effect produced if the lava is also moving, with the long axes of these vesicles parallel to the direction of the lava flow.

Figure 2.18: Vesicular Basalt boulder

- **Amygdules** (also called amygdales) are vesicles which have been filled completely or in part with secondary minerals such as quartz, calcite and zeolites. These have crystallized within the vesicles and other spaces from hot mineral solutions which have flowed through fine cracks within the cooled lava flow.

Figure 2.19: Some weathered zeolite-filled amygdules in an amygdaloidal basalt

2.8 Magmatic Differentiation

The great variety of igneous rocks found within the crust, suggest that it may be possible for several different rock types to be produced from the same magma chamber. One process which can cause such a variety is called **magmatic differentiation**.

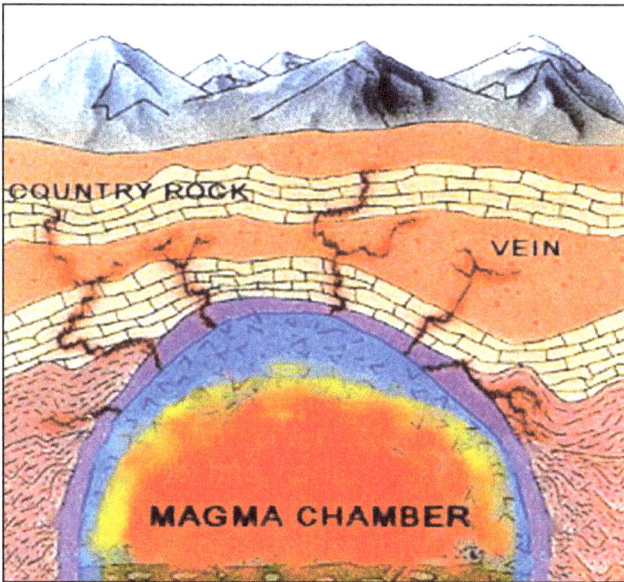

Figure 2.20: A magma chamber penetrating surrounding rocks

Evidence for this process is suggested by the:

- Ability for a single volcano to produce different lavas at different times due to differentiation or separating out of its magma in the chamber supplying the volcano;

- Existence in one area of several different rock types produced by igneous intrusions at successive stages; and

- Existence of layered intrusive complexes which contain different layers of segregated igneous rocks within the one large intrusion or pluton. Such complexes include those of Skaergaard (Greenland), Stillwater (U.S.A.),

Bushveld (South Africa), Sudbury (Canada) and the Giles Complex (Central Australia).

Mechanisms for how different igneous rocks could be formed from the one magma include:

- Gravitational differentiation in which denser minerals (e.g. olivine), crystallizing in the initial stages of cooling (see Bowen's Reaction Series) fall down through the liquid magma and settle out as layers near the bottom of the magma chamber. Sometimes these minerals may not settle at the bottom of the chamber, but may be held up by currents in the fluid. By this method of differential crystallization and density settling, it is possible for a basaltic magma to produce olivine gabbro, at the bottom, with gabbro, diorite and granite at the top, in the one pluton such as a batholith. Valuable mineral concentrates such as chromite (iron-chromium oxide) and magnetite (iron oxide) may also be formed by gravitational differentiation.

- Filter pressing occurs when earth movements, which are often associated with igneous intrusions, squeeze out the liquid part of the magma, leaving any solid crystals behind. This liquid may be removed to a new area where crystallization and more filter-pressing may occur before the melt finally hardens.

- Flow sorting may result if, in a liquid-solid mush, some of the crystals have a definite shape such as prisms which will align themselves and fit together. Often, the walls of vertical igneous intrusions may be lined with a layer of prismatic feldspar crystals.

- Separation of immiscible liquids in much the same way as cream separates from milk, some magmatic liquids may separate simply because they do not mix. After separation they may then crystallize giving different igneous rocks.

- Volatile differentiation refers to the gases dissolved in the liquid magma. These may flow to various parts of the magma chamber, reacting with the different solids or producing crystals themselves, especially in any spaces above the magma chamber. In a pluton, the last minerals to crystallize are often orthoclase, muscovite and quartz. These minerals are often found in pegmatites, igneous rocks which have very large crystals (often metres long) which have been slowly formed in the large spaces left by the removal of the volatile materials.

After the main minerals have crystallized, the residue fluids of hot water, gases and dissolved minerals, called hydrothermal fluids, may escape, filling up veins, fissures and porous rock surrounding the pluton with many late-stage minerals (e.g. quartz) as well as valuable minerals of lead and zinc and possibly silver and gold.

2.9 Some Common Igneous Rocks

Extrusive igneous rocks are formed on the surface by volcanic activity and include:

- Basalt is the most common of all extrusive rocks, being a product of extensive lava flows on land as flood basalts (e.g. the Deccan of India), the material of mid-plate volcanoes (e.g. Hawaii), and the rock of the oceanic crust. It is uniformly black and dull in lustre, but may show some small rectangular crystals of plagioclase when coarse and vesicles and amygdules when fine-grained. Scoria is basalt which is almost completely covered with vesicles due to the high gas content of the original lava. In Hawai'i, the twisted form which results from very fluid lava is called Pahoehoe (from the Hawaiian verb hoe, to paddle since paddles make swirls in the water), whilst the more viscous, slow-flowing solid lava is called A'a (from the Hawaiian word meaning to burn). These terms have been adopted for use worldwide.

Figure 2.21: Basalt, Hawai'i

Figure 2.22: Scoria, Hawai'i

Figure 2.23: Pahoehoe lava near Hilo, Hawaii

Figure 2.24: A'a lava near Kilauea volcano, Hawaii

- Pumice is a frothy-looking rock of granitic composition which has been formed as foam, usually form undersea eruptions. It floats on water because of the large number of vesicles which gives it a density less than that of water.

Figure 2.25: Pumice, Queensland, which has floated in from undersea volcanism in the western Pacific

- Andesite is the aphanitic form of diorite, and is named after the Andes Mountains where it is common. Andesitic volcanoes are usually very explosive due to the volatiles within the magma. The rock varies greatly in colour but it is usually dark green or black with some bigger crystals of rectangular andesine feldspar (a form of plagioclase).

Figure 2.26: Mt. Tungurahua (from the Chechua language – Throat of Fire) - an andesitic volcano in the Andes of Ecuador

Figure 2.27: A hand specimen of a coarse andesite showing the blocky, green andesine feldspar crystals

- Trachyte is a light grey, fine-grained rock with uniform black specks of biotite and amphibole set within a background of feldspars and quartz.

Figure 2.28: Hand specimen of trachyte showing black specks of biotite and in thin section (x 25 cross polars)

- Rhyolite is a very fine textured and light-coloured rock which, like trachyte, may show flow lines. Colours are usually a light pink to light grey. It has the same chemistry as granite.

Figure 2.29: Rhyolite showing distinctive flow lines

- Volcanic glasses are very vitreous (i.e. glassy) rocks having chemistry similar to granite and rhyolite. Obsidian is a very smooth glass which is black or dark brown in colour and has a conchoidal fracture. Pitchstone is a rougher version with a lighter brown to black colour.

Figure 2.30: Brown obsidian, from south eastern Queensland, Australia, showing conchoidal (shell-like) fracture

Figure 2.31: Arrowheads made from obsidian

- Volcanic **breccia** (Italian: *brecce* - broken) is a pyroclastic rock made up of irregular and poorly sorted angular fragments blasted out as coarse fragments and then compacted and cemented together.

Figure 2.32: Weathered volcanic breccia showing angular broken fragments

Intrusive Igneous Rocks are formed deep underground and so usually have large, well-formed crystals. They are exposed at the surface through uplift and erosion:

- Granite is a very common pink-coloured rock, often with larger blocky crystals of orthoclase feldspar (sometimes almost with porphyritic texture), more rounded grey quartz, some blocky white to cream plagioclase feldspar, green-black hornblende and small sheets of very shiny black biotite mica. Muscovite mica may also be present but this is difficult to see in hand specimen. There are a great variety of colours in granites, ranging from the well-known pink to white (adamellite) and green granites.

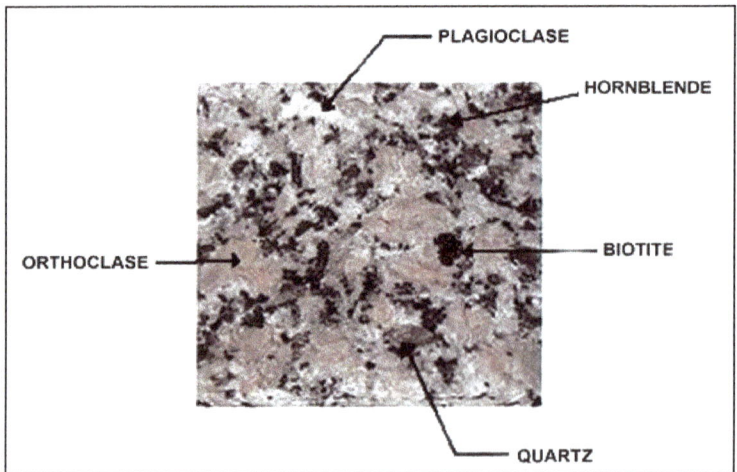

Figure 2.33: A slice of fresh granite showing its main minerals

- Gabbro is a coarse (crystals over 5 mm) rock like granite but darker, with grey plagioclase, green-black pyroxene (e.g. augite) and some dark green olivine. Crystals of grey plagioclase may show beautiful, green-blue **schiller iridescence** when moved in the light.

Figure 2.34: Gabbro showing schiller iridescence in the plagioclase

- Porphyries are a group of acid to intermediate rocks having two major sizes of crystals. Large crystals called phenocrysts form first from the magma during a slower rate of crystallization, then the smaller crystals of the groundmass form very quickly as the residual liquid cools. These rocks are usually named after the main phenocryst e.g. quartz porphyry - a pink rock with large glassy quartz crystals some of which may show hexagonal cross-sections, and hornblende porphyry - a grey to green rock with prominent shiny black rectangles or laths of hornblende: and andesine porphyry having large green phenocrysts of andesine.

Figure 2.33: A quartz porphyry showing well-shaped quartz phenocrysts in a light grey felsic groundmass

- Pegmatites are very coarse rocks with crystals often in tens of centimetres and sometimes over a metre in length. Granite pegmatite is the most common, and this consists of blocky orthoclase, glassy quartz, slabs of muscovite mica and long crystals of green-yellow beryl or black tourmaline.

Figure 2.34: A pegmatite containing mostly of cream orthoclase, some glassy grey quartz and sheets of muscovite

Chapter 3: Sedimentary Rocks

3.1 Introduction

Almost as soon as igneous rocks have been exposed to water and the gases of the atmosphere near or on the Earth's surface, they begin to weather and start the next process of the great recycling system called the **rock cycle**.

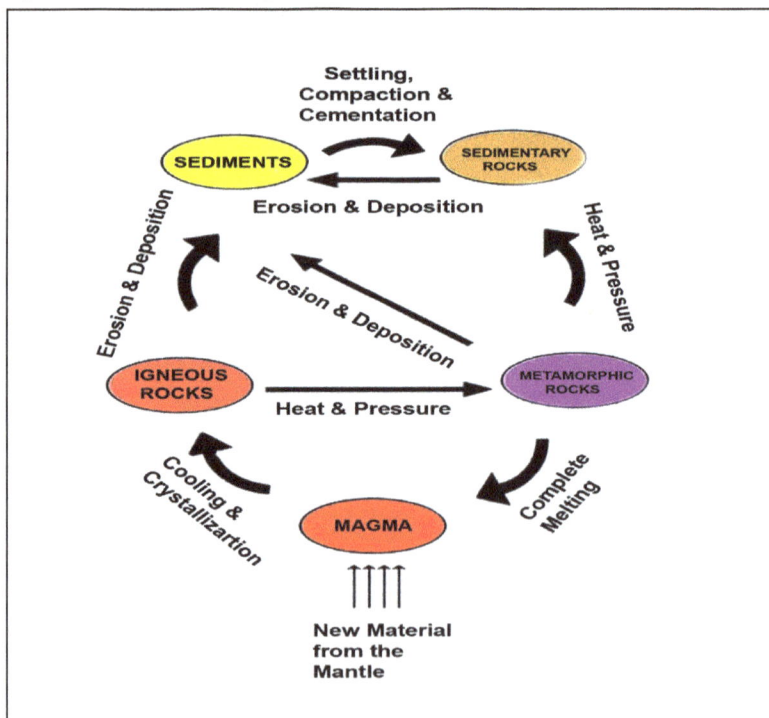

Figure 3.1: The rock cycle

Part of the rock cycle involves the erosion of existing rocks and soil, and the deposition of this sediment into low-lying basins, as the medium of transport such as water, ice or wind, loses energy. This sediment is usually put down in layers or **strata**, with the youngest layers on top of older layers – this is the **Law of Superposition.**

For example, in a mountainous area of igneous rock, weathering by water, gases of the air and changes in temperature, breaks down the rock into softer, smaller pieces. Erosion by gravity, moving water, ice or wind will then carry the particles to lower places where the transport medium loses most of its energy, and the particles are deposited as sediment. There is a recycling of this entire process as parts of the Earth's surface are pushed up as broad vertical movements or during mountain building.

Figure 3.2: Layers of sandstones and shales of the Cliffs of Moher, Ireland

3.2 Sedimentary Rock Formation

The formation of rock is called **lithification** and for sedimentary rocks this process involves weathering, transportation, deposition and compaction with final cementation:

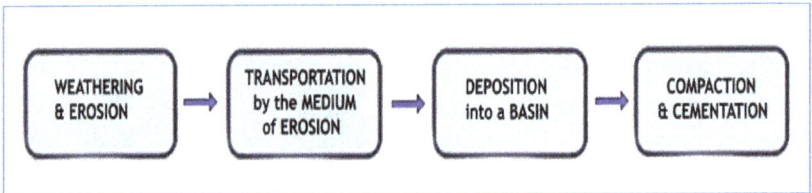

Figure 3.3: The process of lithification

Erosion of rock produced pieces of sediment, or **clasts** of different size and shape with the size depending upon the speed of the transporting medium, and shape depending upon how the rock clasts were transported:

- Fast-moving water in the headwater tract of rivers, and with storm waves on coastlines, produced large, semi-rounded clasts such as boulders and pebbles because of the general rolling action of the moving water.

- Medium-flow water in the middle tract of rivers and the most common ocean waves produces clasts such as small and rounded sands.

- Slow-moving water in the coastal tract of rivers in deltas and estuaries, produce clasts which were very fine and well-rounded, such as silts and muds. Even finer material such as clays usually only settle in deeper, and less disturbed basins such as deep lakes and ocean basins.

- Glaciers usually produce poorly-sorted clasts, i.e. a variety of shapes and sizes all mixed together such as glacial till. This is done by the grader effect of the glacier pushing and grinding the rock (called moraine) at its edges and base. Meltwater streams emerging from the glacier will have more rounded clasts due to the rolling action of the water and will also deposit clasts of different sizes depending upon the speed of the water.

- Wind-eroded particles in deserts will produce clasts which may be medium to large in size due to the abrasion of the sand on previously broken rock on desert pavements. These are called **ventifacts.**

- Wind-blown sands may be less rounded than water-borne sands and are deposited as dunes in deserts or behind beaches. In the deserts, streams on fan-like wedges of sediment may further re-work sediment into fine silt. Though not strictly clasts, crystals of mineral salts also form significant layers at the end of these streams on desert flats or in playas.

Deposition of sediment requires a low-energy, lower relief environment, or an obstruction to offer similar conditions, for the speed of the medium of transportation to be reduced enough for the sediment to stop its motion and settle. In general terms, the faster the medium of transportation, the bigger the clasts which can be moved and when this medium slows down sufficiently then clasts of different sizes will be deposited in order of size with the larger clasts settling first and the very fine material last of all.

In addition, the shape of the clast will also determine its rate of transportation and deposition. This also concerns the **sphericity** and **roundness** of the particles. Those clasts which are, flat or oblong or are irregular in shape have a slower rate of transportation and deposition than those particles which are almost spherical.

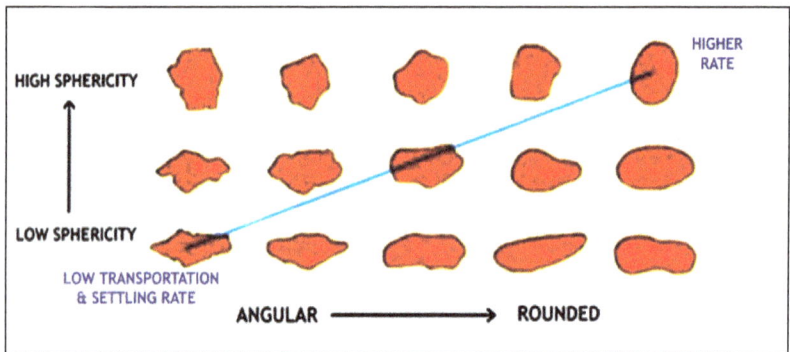

Figure 3.4: Particle shape as a combination of sphericity and roundness

As more sediment is deposited on top of the older sediment, it is pressed together and water is removed. As this occurs, dissolved minerals precipitate between the interlocked grains and act as cement. Silica (quartz), iron oxides (e.g. haematite) and carbonates (e.g. calcite) are common cements found in sedimentary rocks. As well as the cement and grains of the rock, its matrix, or powder between the grains, is also important in the conversion of sediment into sedimentary rocks.

The nature of the grains may also indicate the source of the fragments. For example, rounded, spherical grains of similar size are well-sorted because they have been constantly rolled around and abraded within a water environment. Angular and poorly-sorted fragments may have come from an environment where the clasts have been pushed such as at the edges of a glacier, or fallen together such as at the base of a cliff. Poorly-sorted, un-weathered grains may indicate rapid burial, such as in rocks produced by landslides or by underwater turbidity currents which carry sediment off a collapsed coastline. With normal water and wind erosion, the longer the transportation, the better the **sorting** of the clasts upon deposition.

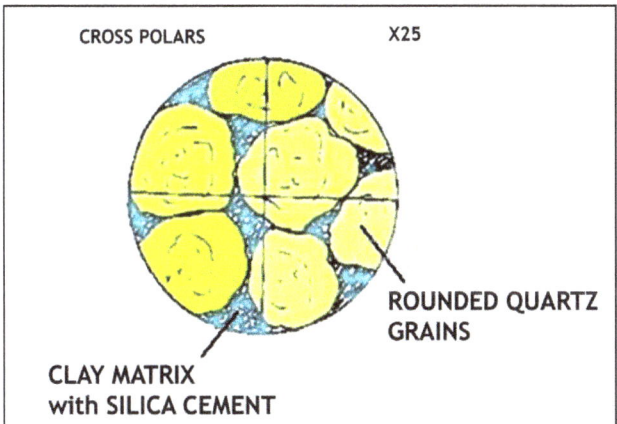

Figure 3.5: A thin - section of a fine sandstone showing clay matrix and cement

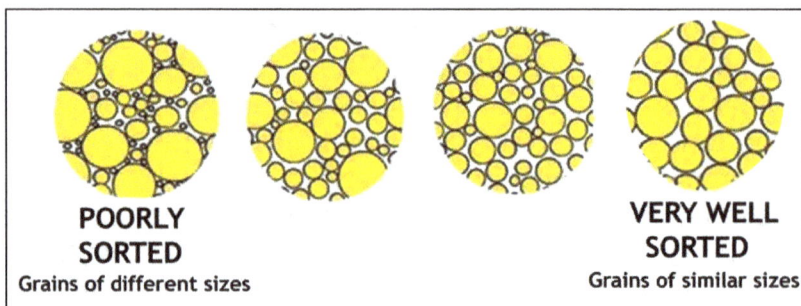

| POORLY
SORTED
Grains of different sizes | | | VERY WELL
SORTED
Grains of similar sizes |

Figure 3.6: Sorting of clasts within a rock

3.3 Classification of Sedimentary Rocks

There are many ways of grouping and classifying sedimentary rocks. This may include classification by their:

- Environment of deposition. For example, **fluvial** sedimentary rocks are deposited under freshwater streams, **lacustrine** under lakes, **paludal** sediments are deposited under swamps and marshes and **aeolian** sediments are deposited by wind. There are also glacial and desert sedimentary environments. These environments are often grouped more generally as **terrestrial** environments to distinguish them from marine environments and the term can also be used to describe those specific environments in which loose material or regolith are deposited on bare land surfaces such as the base of cliffs due to landslides.

- Composition in terms of their most common chemical or mineral constituents e.g.

  - Siliciclastic sedimentary rocks are mainly composed of silicate minerals with much silica (quartz). They are very resistant to weathering e.g. sandstones.

  - Siliceous sedimentary rocks are almost entirely composed of silica which have been deposited or precipitation from a silica solution, such as chert and flint.

  - Carbonate sedimentary rocks are composed of calcium and/or magnesium carbonate such as limestone or dolomite.

  - Evaporite sedimentary rocks are composed of minerals formed from the evaporation of water. The most common evaporite minerals are carbonates such as travertine, rock salt, and gypsum.

  - Organic sedimentary rocks have significant amounts of organic material and coal, oil shale, coquina (shells) and the organic-rich source rocks for oil and natural gas.

  - Iron-rich sedimentary rocks are composed of greater than 15% iron with the most common forms being banded iron formations and ironstones.

  - Phosphatic sedimentary rocks are composed of phosphate minerals and include deposits of phosphate nodules, bone beds, and phosphatic mud rocks.

- Particle size, which is probably the best classification, as it generally eliminates many of the ambiguities of the previous two classifications in which there is considerable overlap. In this classification, sedimentary rocks are grouped according to either having observable particles or clasts as **clastic** sedimentary rocks, or having no visible particles as **non-clastic** sedimentary rocks. These major groups can be further classified according to other parameters such as how they were formed.

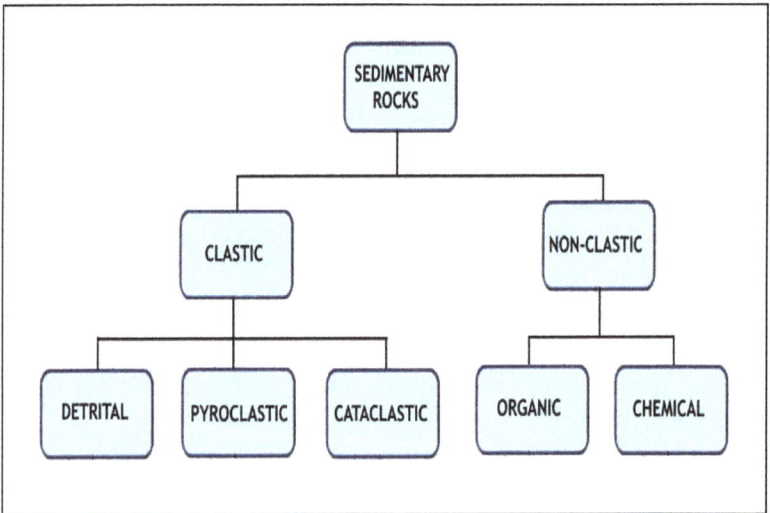

Figure 3.7: A simple classification of sedimentary rocks

3.4 Clastic Sedimentary Rocks

These are rocks which contain particles derived from other sources by weathering and erosion. They include pyroclastic rocks produced from explosive volcanoes, turbidites from below the ocean, and rocks formed from fault movement and landslides. **Cataclastic rocks** belong to a small group which may be found within limestone of karst sequences where they have been formed from the collapse of large caves.

For the main group of clastic rocks, the size of the particles is not only important for reference purposes, but also as an indication of the mode of transportation and deposition. Whilst there have been several scales or lists developed for the classification by particle size, that of J. A. Udden (Swedish-American: 1859-1932) of 1898 and modified by C.K Wentworth (American: 1891-1969) in 1922 is still useful in describing clastic sedimentary rocks.

For practical purposes, standard sieves are used to determine the percentage grain size for any rock or sediment. The sample rock is weighed and then lightly broken up into its component grains without damage to them. The resulting powder is then passed through a stack of sieves with largest on the top and descending order to the smallest at the bottom. This stack is then mechanically shaken and then separated into its individual sieves. The content of each sieve is then weighed and the percentage of grain sizes in the rock can be estimated from the original whole-rock weight. From these percentages, a **modal analysis** of grain size for that sediment or rock can be found and represented in graphical form:

Grain Size	Sieve Scale Φ	Aggregate Name	Alternative Name	Rock Names
> 256 mm	−8	boulder		
64-256 mm	−6 to −8	cobble		
32-64 mm	−5 to −6	very coarse gravel	pebble	
16-32 mm	−4 to −5	coarse gravel	pebble	conglomerates, agglomerates & breccias
8-16 mm	−3 to −4	medium gravel	pebble	
4-8 mm	−2 to −3	fine gravel	pebble	
2-4 mm	−1 to −2	very fine gravel	granule	
1-2 mm	0 to −1	very coarse sand		
½-1 mm	1 to 0	coarse sand		
¼-½ mm	2 to 1	medium sand		sandstones & arkoses
125-250 µm	3 to 2	fine sand		
62.5-125 µm	4 to 3	very fine sand		
3.90625-62.5 µm	8 to 4	silt	mud	siltstones
< 3.90625 µm	> 8	clay	mud	claystones
< 1 µm	>10	colloid	mud	

Table 3.1: The Wentworth-Udden Scale of particle sizes

(The phi symbol relates to the Krumbein phi Sieve Scale which is -log2 diameter of grain / reference diameter of 1mm. This is useful in sieve size calculations. The symbol µm is the unit, the micrometre, which is one millionth of a metre)

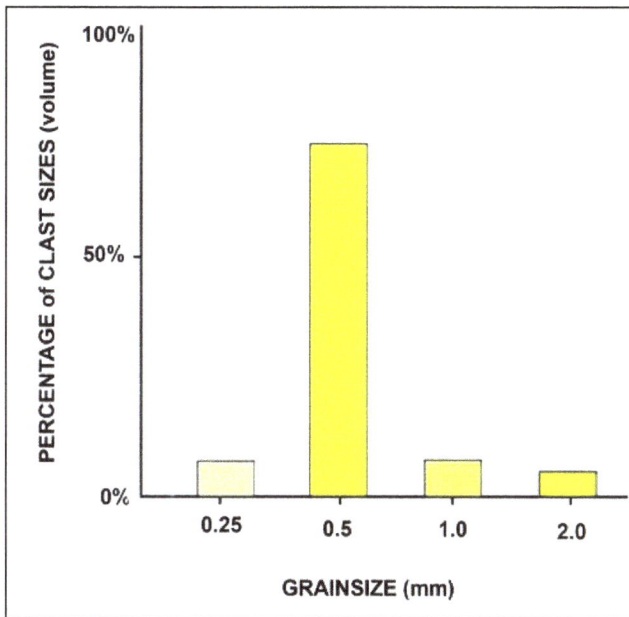

Figure 3.8: Modal analysis of a well-sorted, medium to coarse grained sandstone

In this graph, it can be seen that the majority of grains are in the 0.5 mm range, so from the Wentworth-Udden Scale, it would be classified as a medium to coarse sandstone. The terms **rudites**, **arenites** and **lutites** are sometimes used as field terms but are not often used nowadays and generally refer to clastic rocks which are coarse grained (e.g. conglomerates), medium grained (sandstone) or fine grained (mud and siltstones) respectively. These terms are useful when classifying transitional rocks such as conglomeratic sandstones or sandy shales.

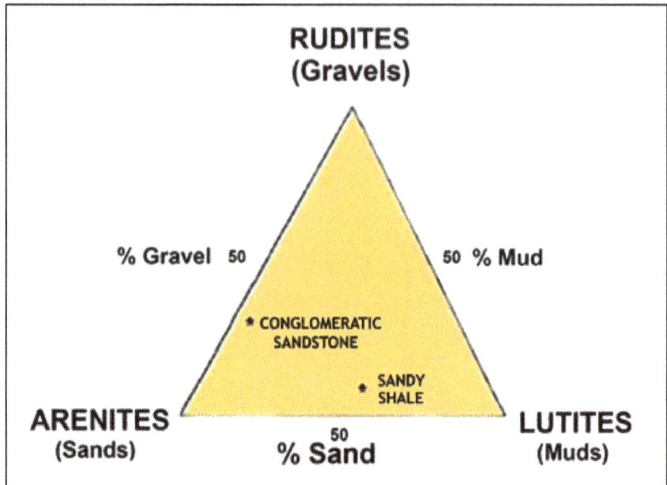

RUDITES (Gravels)

% Gravel 50   50 % Mud

* CONGLOMERATIC SANDSTONE

* SANDY SHALE

ARENITES (Sands)   % Sand 50   **LUTITES (Muds)**

Figure 3.9: A triangular graph (3 components) for clastic sedimentary rocks

The term mudstone is often used as a field term for any fine-grained rock which could consist of clasts in the official silt to colloid range of the Wentworth-Udden Scale. Often the term shale is used when referring to layered mudstones. True siltstones will be in layers like most sedimentary rocks, but these are often very massive, that is, layer which are very thick and show few distinctive internal layers. Similarly, the terms sandstone and conglomerate are field terms with each having a wide range of clast size.

Some common examples of clastic sedimentary rocks are given below. It is worth noting their features as many rocks found in the field may be of great age and the current, local environment will have little or no relationship to the environment in which these rocks were formed. For example, in parts of dry, inland Australia, it is possible to find fossiliferous limestone several hundred metres above sea-level, indicating that this region was once under a warm, shallow sea.

Some of the most common clastic sedimentary rocks are:

- Volcanic breccia is a large-grained pyroclastic rock consisting of a mass of poorly-sorted, angular fragments, of varying composition of volcanic material ejected explosively from a volcanic vent. Often called agglomerate referring to the amalgamation of particles in a large mass rather than clasts cemented together. It may also be classified as an igneous rock because of its volcanic source.

Figure 3.10: Volcanic breccia badly stained by iron oxide

- Sedimentary breccia is a similar rock, but they have a great variety of angular clasts in size and composition and have been produced by a sudden collapse of a rock formation such as in a landslide. Clasts are usually immature, that is, show little signs of much reworking by transportation, having been formed not far from their as the result of purely gravitational erosion. They result from the compaction and cementation of finer material from scree slopes in steep-sided river valleys, at the base of desert plateau, and at the base of sea cliffs. On a larger scale, the total deposit may be wedge-shaped indicating their sudden deposition at the base of a cliff face.

Figure 3.11: Sedimentary breccia showing angular rock fragments

- Conglomerate has large, rounded clasts of various sizes, composition and sorting but clasts are usually greater 2mm in diameter. Usually conglomerates will show poor sorting, indicating sudden deposition from a fast-flowing mountain stream or on a storm beach. Better sorting will occur with longer periods of transportation in a river with a high flow rate or with constant re-working by waves of uniform intensity on a storm beach. The differences between a fluvial and a marine conglomerate may be seen by the presence of freshwater organic remains in the fluvial sediments, such as pieces of fossil woods and carbonized fragments, and in the marine sediments there may the presence of fossils such as shells and imbrication or flat stacking of clasts. There may also be differences in the shape of the whole deposit, for example, a conglomerate bed may show the curved shape of a point bar which was on the bend of river or the crescent shape of a storm beach. Conglomerates formed by glaciers will be poorly sorted and have a great variety in size and consist of the very angular clasts.

Figure 3.12: Hand specimen of a weathered conglomerate. The red colour is due to iron oxides

As well as by environment, conglomerates may be classified simply by their composition. **Oligomictic** conglomerates are composed of clasts of mainly one type of mineral or rock, whereas **polymictic** conglomerates have a great variety of clast types. The homogeneity, or lack of it, is determined by the sources of the clasts.

Figure 3.13: part of a large bed of polymictic conglomerate showing the sub-rounded clasts of quartz and a variety of rock

- Greywacke, named from the German *grauwacke*, for a grey, earthy rock, is also a poorly-sorted rock containing angular to sub-angular clasts of a variety of compositions. A greywacke deposit may also include large, sub-beds or lenses of other rock, such as limestones. It is usually found as large deposits associated with other marine sedimentary rocks, and represents a large under water avalanche caused by a **turbidity current** - a large, downslope flow of sediment and water coming off the continental shelves into the deeper water of ocean floors. It is often more finely-grained than terrestrial conglomerates

Figure 3.14: Greywacke specimen showing poor sorting

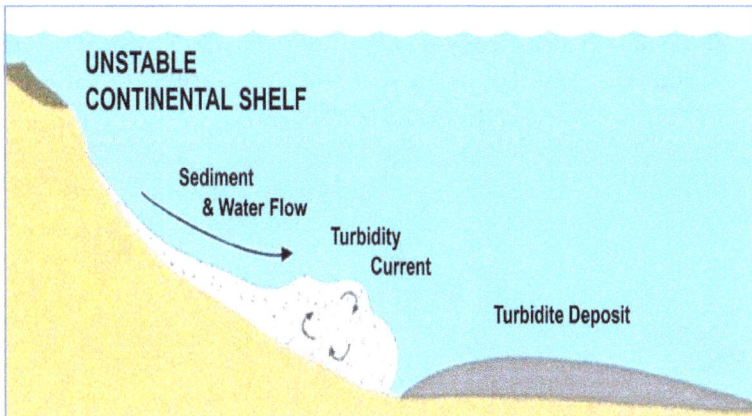

Figure 3.15: Diagram showing a turbidity current

- Orthoquartzite, commonly called sandstone, is comprised of grains of quartz, originally deposited as water or wind-borne sands in rivers, coastal margins and in desert depressions. Wind-borne sands such as in dunes are well-sorted and fine grained, beach sands have medium sorting and size, and river sands have variable sorting and medium to coarse grainsize. Some sandstones from relatively immature river systems may also contain small, whole rock fragments from their source areas. These rock fragments have not yet completely weathered, and the sandstones which contain them are called lithic sandstones. Other sandstones may contain a large percentage (> 25%) of fresh feldspar mineral which also has not had time to be completely weathered. These sandstones are called arkoses, and they too, indicate a relatively close source, perhaps quartz-rich igneous rocks such as granites, or metamorphic rocks rich in felsic minerals such as gneisses. Cement in sandstones and similar clastic sedimentary rocks can be iron oxides, clay, silica or calcite with the latter being common in arkoses and calcareous sandstones.

Figure 3.16: medium to coarse quartz sandstone with average grain size about 0.5 to 1.0 mm

Figure 3.17: close-up of sandstone of the Big Arm on the planet Mars, showing fine to medium rounded grains. (NASA/JPL-Caltech/MSSS photo.)

- Mudstones are rocks which have very fine grains of sedimentary material (usually less than 125 μm). The term mudstone is a general or field name for any fine rock such as a siltstone or claystone. Mudstones which have layers, called **laminae**, are often called shales.

Siltstones are coarser than claystones, and these rocks have very fine grains of mostly very fine quartz and clays, and are usually very well-sorted. They are formed in relatively quiet water such as the lakes, lagoons and swamps of the coastal tract of rivers, deltas and offshore in deeper marine environments where there is little wave action. In places where there was abundant life, they may contain plant or animal fossils which are found between the laminae which once formed the top of the mud layer.

Figure 3.18: A siltstone with iron-oxide staining on top

Figure 3.19: A shale showing fine laminae

Figure 3.20: Strata of fine-grained rocks, probably a shale on Mars in the Gale Crater (NASA/JPL Caltech/MSSS photo.)

Of importance are **varved shales** (Swedish: *Varv* for revolution/layers) which are produced when very fine glacial till is washed out of a glacier and into lakes which may form in the valley beyond their front or snout. This fine till is washed into the lake at each summer melt of the ice, giving a new and separate layer. Counting the layers within varve shales has been used for age determination in geological time estimations.

Figure 3.21: Varve shales in the valley of the Morado Glacier, Chile

Oil shales are another important fine-grained sedimentary rock. They are dark grey-black in colour, and have a significant amount of their pore space between the grains. These pores are filled with the hydrocarbon kerogen and they can be used as a fuel source if they are crushed and the hydrocarbon extracted.

3.5 Non- Clastic Sedimentary Rocks

These are sedimentary rocks which have been formed by chemical precipitation, crystallization from concentrated solutions, or by the accumulation of the remains of living things. They include:

- Chemical sedimentary rocks which are formed from chemical precipitation and settling of the solids when water has evaporated from concentrated mineral solutions or from the cooling of hot solutions allowing the dissolved mineral to crystallize. These rocks are often **monomineralic**, having usually one main mineral with smaller amounts of impurities. Some examples include chert, halite (rock salt), gypsum, anhydrite and chemical limestone.

- Organically-formed sedimentary rocks are those formed from the compaction and cementation of the remains of living things. They include coal from plants, fossiliferous limestone from corals, and coquina from shells and diatomite from small marine plankton.

Some of the most common examples of chemically-formed sedimentary rocks are:

- Chert, flint and chalcedony which all have been deposited as a jelly-like precipitate of quartz. This comes out of solution and is then compacted and hardened into layers in the rock. Chert is a hard, smooth rock with conchoidal fracture and can have a variety of colours such as white, yellow, brown and black. It can also be biological in origin, when the silica is recrystallized from the dissolving of silica shells (tests), of marine plankton and sponge spicules. In both cases, crystals of silica are interlocking and microscopic (**cryptocrystalline**). Chalcedony (from the Latin: *Chalcedonius* derived from the town Chalcedon in Asia Minor) is a very fine cryptocrystalline form of silica, closely resembling the properties of the mineral quartz and has a waxy lustre. Flint is usually a black colour and is often found as nodules or small lenses within softer marine rocks such as limestones.

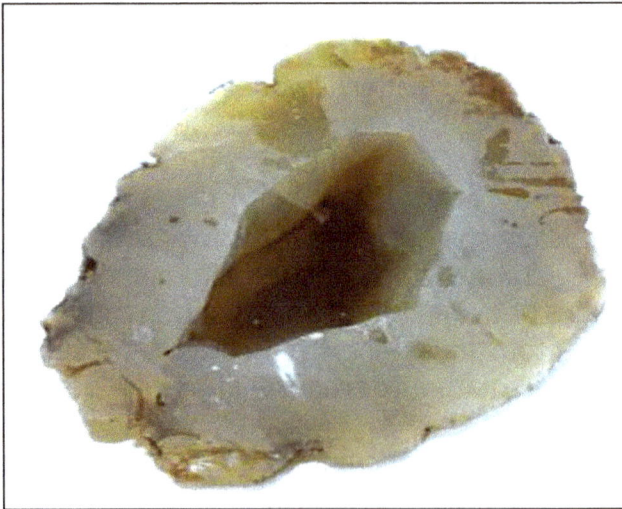

Figure 3.22: a cut cross-section through a large (15 cm) chalcedony geode which precipitated in a cavity.

- Rock salt and is composed almost entirely the mineral halite or sodium chloride (NaCl). Underground salt deposits are initially formed as an evaporite deposit as shallow seas evaporate. This is then buried by more sediment to be squeezed upwards as large underground domes called a **diapir**.

Figure 3.23: Cubic crystals of halite (salt)

Figure 3.24: Inside the ancient salt mine at Hallein, Austria

Anhydrite and gypsum are other deposits which are extensive enough to consider as rock. Anhydrite is composed of calcium sulfate ($CaSO_4$) which and is found as an evaporite deposited along with its **hydrated** form gypsum. Gypsum ($CaSO_4.2H_2O$) has been an important resource for thousands of years. These rocks form as accumulations of natural evaporites from their mineral solutions which have been leached out from other rocks. As solutions, they form in natural basins and then accumulate as the water evaporates. This may occur in desert playa lakes and enclosed seas. They may also be deeply buried, compressed and formed as a rock below ground and then be forced up as a diapir.

- Chemically-formed limestone is usually found in smaller outcrops than organically-formed limestones because they are formed in lakes, on or near sources of calcium ion (Ca^{2+}), such as in the calcium plagioclase (*anorthite*) in the rock basalt. For example, if a lake is formed on an old basalt lava flow, the calcium ions may be leached out of the rock over time, and react with the carbon dioxide dissolved in the water of the lake producing calcium carbonate. This settles, becomes compressed and lithifies as a chemical limestone deposit. The rock is usually soft and rarely has fossils.

Figure 3.25: Chemically-deposited limestone, Limestone Park, Ipswich, Australia

Figure 3.26: An outcrop of chemically-deposited limestone, Limestone Park, Ipswich, Australia

- Travertine is another form of chemically-deposited limestone. It is formed from hot springs and geysers around which the rock is deposited from the hot water solutions containing the calcium carbonate.

Figure 3.27: Travertine deposits around the Pohutu Geyser, Rotorua, New Zealand.

Some of the most common examples of organically-formed sedimentary rocks are:

- Organically-formed limestone which is formed from the dead and compressed remains of animals of the extensive coral reefs that form in the clear, warm waters of tropical seas. It consists mostly of the mineral calcite (calcium carbonate $CaCO_3$), from the skeletal remains of corals, with some also coming from foraminifera (or forams, small, single-celled animals), crinoids and molluscs. As the coral reefs are built by successive layers of new generation corals, the dead material is compacted and becomes rock. This may then be uplifted out of the sea by earth movements. Limestone is usually a blue-grey colour, and forms extensive outcrops over wide areas (called karst).

Figure 3.28: Fossiliferous limestone showing small solitary corals

Figure 3.29: Crystalline limestone

Figure 3.30: Carlotta's Arch – part of the massive limestone deposit at Jenolan Caves, New South Wales, Australia.

- Oolitic limestone contains small **oolites** (Ancient Greek: *òoion* for egg), round spheres showing concentric internal structure. They are formed on calcareous beaches rich in carbonate material such as shells and coral, and with a slope sufficient to allow a regular, but gentle washing of the sand or small coral grains. The dissolved calcium carbonate in the water successively coats the grain with precipitative layers, giving them their distinctive shape.

Figures 3.31: Oolitic limestone in hand specimen showing oolites

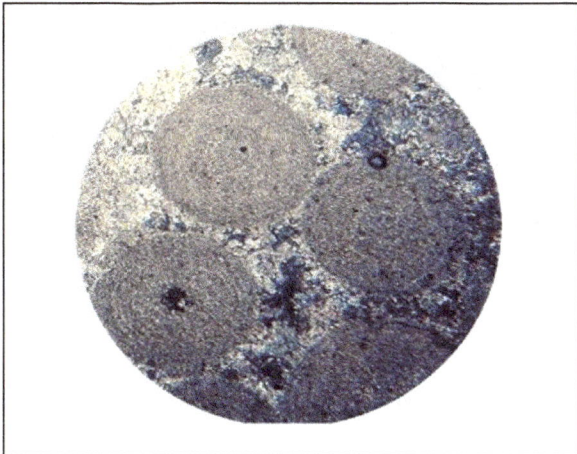

Figure 3.32: Thin-section of oolitic limestone with cross polars showing the concentric internal structure of the oolites and the bright calcite cement

- Dolostone forms where calcite mineral has reacted with high concentrations of magnesium ions (Mg^{2+}), such as found in some organic-rich marine muds in similar environments to those which form limestones. This rock mostly contains the mineral dolomite (calcium-magnesium carbonate – Ca Mg $(CO_3)_2$) and there is still some doubt as to the exact mechanism of how dolostone forms. Fossils in dolostone are few, and then poorly preserved. This rock forms huge outcrops such as The Dolomites Alps in northern Italy.

Figure 3.33: Part of the Dolomites, a mountain range on the Austrian-Italian Border.

- Coquina from the Spanish for cockle shell, is a soft, poorly-cemented, sedimentary rock made entirely of compressed shells of marine animals found in high-energy environments of the seashore. The mineral composition is therefore similar to limestone, being mostly calcite from the shells.

Figure 3.34: Coquina made from compacted shell fragments (Photo: Wiki Commons)

- Diatomite consists of the accumulated tests or silica shells of microscopic diatoms, singled-celled marine algae which form large deposits when they die and sink to the sea floor. The rock is soft, light in weight, usually white to light brown in colour, is easily broken (i.e. is **friable**) and generally featureless. It is mined as diatomaceous earth and is used in filters and as an abrasive cleaner.

Figure 3.35: Photomicrograph of marine diatoms
(Photo: USGS)

- Coal is formed from the anaerobic (without air) breakdown of large volumes of plant matter which has fallen into the quiet waters of freshwater swamps (paludal environments) and lakes (lacustrine environments). Coalification is the process which turns compressed vegetation into coals with the loss of water (H_2O), carbon dioxide (CO_2) and methane (CH_4). The extent and rate of this process depends upon:

  - type of original vegetation

  - depths of burial

  - temperature and pressure at these depths

  - length of time

With time, deep burial with the associated temperature and pressure, compressed vegetation begins to transform into various ranks of coal. Each rank develops into the next and more superior variety with further compression and coalification. The main ranks of coal are:

- Peat is often considered only a precursor to coal as it consists of considerable vegetable matter with little coalification, a high percentage of water and silica and clay as soil. It is found extensively in parts of northern Europe with large areas of marsh and bog. It has, however, been a source of poor fuel for centuries and is still dug for local consumption in Ireland and Finland.

Figure 3.36: Peat showing the fibrous nature of this rank of coal (Photo: USGS)

- Lignite or brown coal is the lowest true rank of coal, having about 25-35% Carbon but still high in moisture content and amount of silica (or ash). It is useful for generating heat in power stations.

Figure 3.37: Lignite or brown coal showing woody texture at right

- Sub-bituminous coal has 35-45% carbon content, about 15-30% moisture and considerable ash content. These coals are not suitable for making coke, the solid fuel made from heating coal without air, but they are useful for heating in steam boilers. Together with lignite, this rank makes up the largest reserve of the world's coal.

- Bituminous coal is the typical hard, black coal used in making coke for industry and most high-grade steam-generation boilers. It contains bright and dull bands and contains 45-86% carbon with some moisture and ash.

Figure 3.38: A banded bituminous coal showing bright bands

- Anthracite is a very hard, often glossy in lustre and may show conchoidal fracture as it is very brittle. Has 86-98% carbon and is considered as a transition towards pure carbon as graphite after further change by heat and pressure.

Figure 3.39: Anthracite coal showing conchoidal fracture

About three metres of compressed vegetation will eventually form one metre of coal which is formed into layers called seams. These are contained within horizontal sequences of sedimentary rocks, which are usually freshwater shales, sandstones and conglomerates. These are often called coal measures. Seams can be many metres thick, and the area of the seam may be several hundred square kilometres. These coal measures may contain several seams, representing different periods of deposition. Much of the world's coal was formed in the Carboniferous period (about 359.2 to 299 million years ago), although considerable coal formation occurred in more recent times in the Permian (299 to 251 million years ago) and Triassic Periods (251 million and 199 million years ago) in the southern hemisphere. More detail on coal and coal mining is given in the companion book RICHES FROM THE EARTH.

Figure 3.40: Strata showing a sequence of shales and sandstones at top and a coal seam at base - from the Late Triassic Ipswich Coal Measures, Brisbane, Australia.

3.6 Fluids in Sedimentary Rocks

Because they are made up of particles, clastic sedimentary rocks have various degrees of porosity due to the many open spaces, or pores, between the grains. They also have some permeability, meaning that they allow water to pass through them to some degree. Non-clastic sedimentary rocks and even some igneous rocks such as basalt and granite can also be porous and permeable, if they contain many cracks, joints and bedding planes which can hold water or allow it to pass.

However, a rock can be porous but not permeable (or impermeable) if the pores spaces are not connected.

Porosity (Φ) is the ability of a rock to hold fluids such as water, mineral solutions, oil and natural gases, because of the many pores or interstitial spaces between its grains. The porosity of a rock depends upon:

- Rock type – as it is not the size of grains but the space between them which is the important factor. This is called intergranular porosity. For example, a mudstone will have a greater porosity than a conglomerate because there are many more but smaller, pore spaces within the finer-grained mudstone.

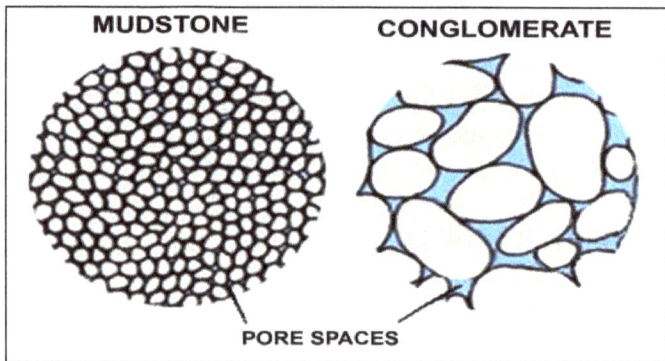

Figure 3.41: A diagram showing intergranular porosity

- Sorting of grains – well-sorted rocks have greater porosity than poorly-sorted rocks because in these rocks, the smaller grains will fill up many of the larger pore spaces.

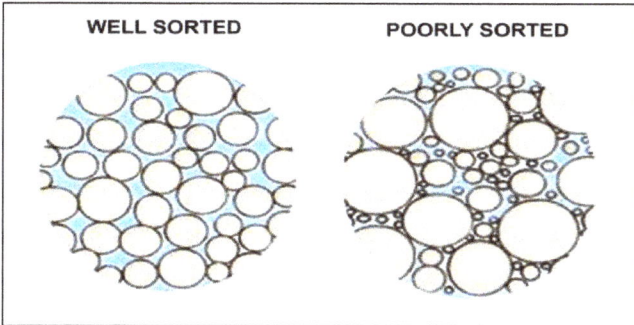

Figure 3.42: Diagram showing sorting of grains

- Cementation – which usually decreases the porosity of the rock by filling the voids between the grains with cementing material.

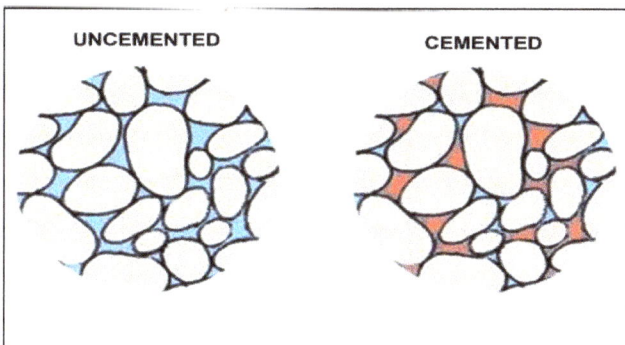

Figure 3.43: The effect of cementation reducing porosity

- Shape of the grains – grains which have good roundness and sphericity give a greater porosity than those which are more irregular in shape because the irregularly-shaped grains can interlock with each other reducing pore space.

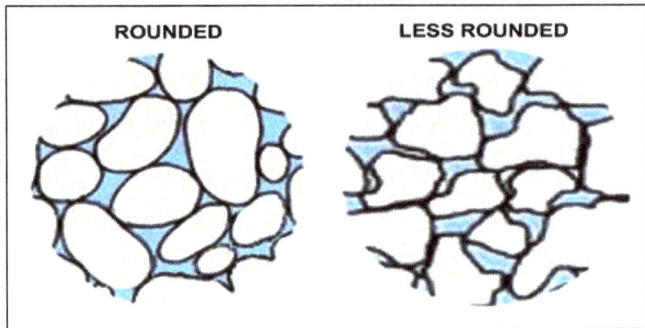

Figure 3.44: Shape of grains

- Compaction – decreases the pore spaces as the grains are rearranged and pushed together

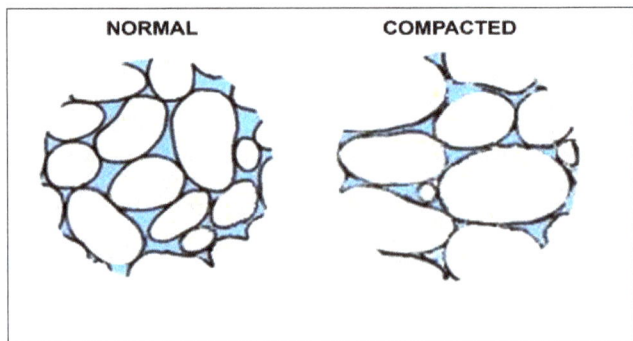

Figure 3.45: Effect of compaction

- Amount of fracturing – increases the porosity by providing additional space caused by the cracks in the rock (fracture porosity). This also allows fluids to move through the rock more effectively.

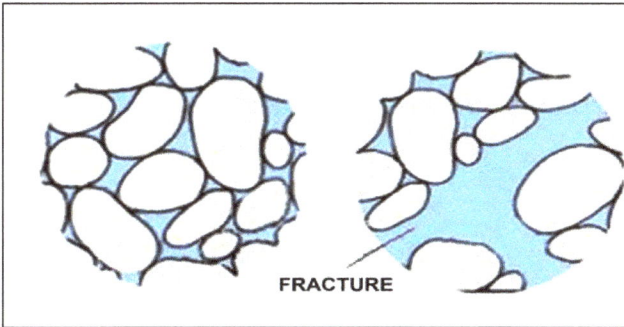

Figure 3.46: Fracture and pore space

Some porosity values from several sources are given below as a rough guide to the ability of different rocks to hold fluids but these will vary greatly with the amount of fracturing and weathering that the rock has undergone.

In scientific terms, porosity is a measure of the total volume of the spaces between grains and is given as a percentage.

$$\Phi = \frac{V_V}{V_T} \times 100$$

Where V_V = volume of the pores; and
         V_T = total volume of the rock.

Permeability (k) is the ability of a rock to allow the passage of fluids through it, either because the rock is fractured or has inter-connected pore spaces or both. Rocks can be porous but not permeable simply because their pore spaces are not connected and the rock is intact and not fractured. Permeability measurements can be very complicated because the rate of flow of a fluid through a particular rock depends upon:

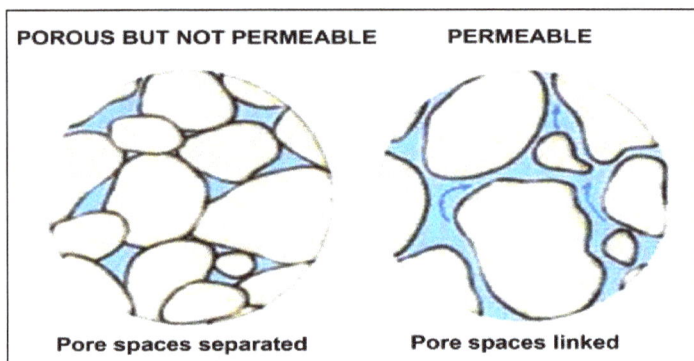

Figure 3.47: Permeability and Porosity

- Rock type – conglomerates and sandstones have high permeability because of their large, connected pore spaces. Basalts and some limestones also can have good permeability because of their high degree of jointing and fractures. Siltstones and mudstones usually have low permeability because their smaller grain size limits their connectivity. Non-fractured granites and other crystalline rocks simply do not have connected pore spaces or fractures to conduct fluids.

- Size of pore spaces – will determine how much fluid will flow through the rock, provided the pores are linked. This contributes to the overall cross-sectional area available for the passage of the fluid through of the rock unit carrying it.

- Viscosity of the fluid, or its ability to flow, concerns the frictional effect which the rock has on the fluid or the inherent nature of the fluid as a sticky substance. Gases, such as methane, will flow better than water which will flow better than oil. **Viscosity** and therefore flow rate, also depends upon the temperature of the rock.

- Hydrostatic pressure or the force which will push the fluids through the rock. This is due to the head, or height difference between where water may enter the permeable rock, such as in a mountain range receiving much rainfall, and where it will flow out at some vertical distance below such as at a spring in a valley. Sometimes, the head might be due to the buoyancy of the gas, water and oil moving upwards through permeable rock layers.

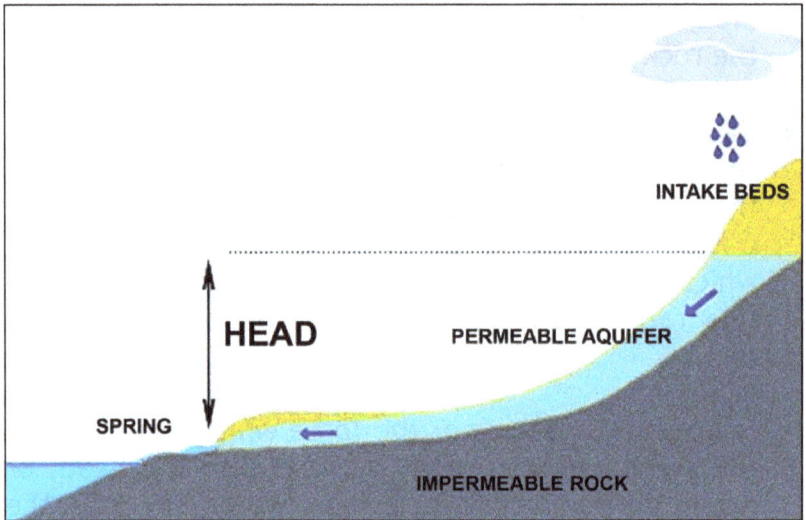

Figure 3.48: Diagram showing the concept of hydrostatic head

Groundwater systems of permeable and impermeable rock layers play a major role in the hydrological (or water) cycle. This is natural recycling of water. The movement of water in the ground through permeable rock units is called hydrogeology or groundwater hydrology.

Permeable rock units which carry water are called **aquifers** and those which do not are called **aquicludes**. Conglomerates, sandstones, fractured rocks such as limestone and unconsolidated sands and gravels are good aquifers whereas clays, claystones, siltstones and non-fractured crystalline rocks are good aquicludes. If a rock has low permeability, but will still carry water, such as very clay-rich sandstone, it is called an **aquitard**. Together, these rocks form an **artesian system** and many countries rely on artesian basins as a source of useable water. Water from precipitation is infiltrated at the recharge zone and is carried down into the permeable rocks of the aquifers into an artesian basin. If this water has been trapped between two layers of impermeable rock as a confined aquifer, then drilling into it may result in an **artesian bore**. Here the water comes out to the surface from very deep underground, often from thousands of metres below, under its own pressure and it is usually hot, charged with minerals and sometimes is accompanied by natural gas. In some parts of the world, such as in the Great Artesian Basin of Australia, the most regular supply of water comes from these artesian bores, however it is only suitable for stock animals.

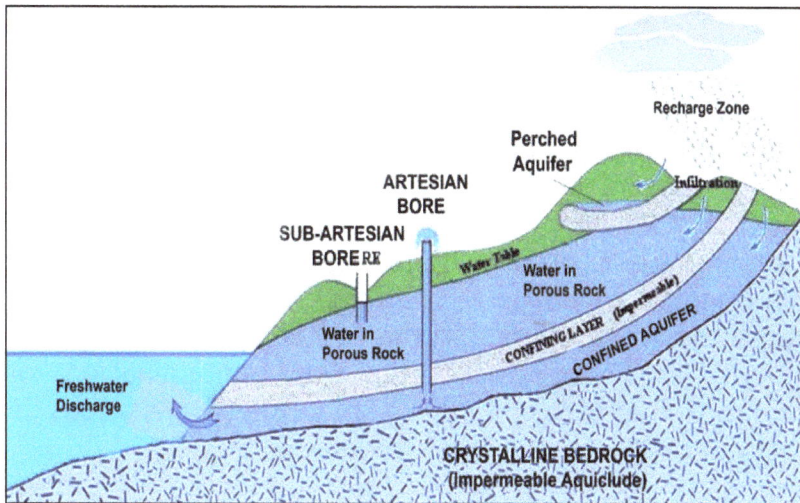

Figure 4.49: Diagram showing an underground artesian system

Where the water is not confined, and simply percolates through permeable rock below, wells or **sub-artesian** bores may be drilled to obtain it and the water has to be pumped to the surface. Recharging these aquifers depends upon regular precipitation in the recharge zone which may be thousands of kilometres from the bore and in an entirely different climatic environment. Water from the recharge zone may be travelling at only a few metres per year so that much of the water coming from the bores today is many thousands of years old. For example, in the Great Artesian Basin of Australia, recharge occurs in the wetter regions along the eastern coastal mountains of the Great Dividing Range which forms a rain shadow for the inland regions serviced by these bores. In the United States, the Edwards Aquifer is one of the major sources of water for agriculture and industry in the state of Texas.

In places where extensive use is made of artesian and sub-artesian water for stock and agriculture there is always the danger that use will exceed recharge rate and levels in bores drop or the bore dries up entirely. This can happen when there are long periods of drought in both the areas of use and the recharge zone. Moreover, there is always the threat of contamination of groundwater through mining and use of pesticides in agriculture.

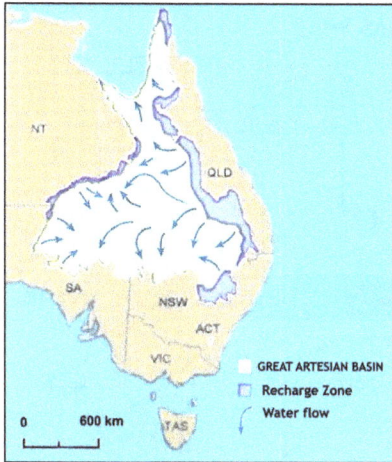

Figure 3.50: The Great Australian Basin of north eastern Australia

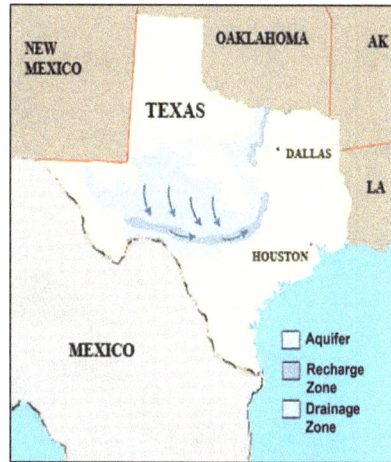

Figure 3.51: The Edwards Aquifer of Texas, USA

Petroleum as a mixture of crude oils and natural gases is sometimes found within suitable sedimentary rocks. These **hydrocarbons** have been formed from vast quantities of microscopic dead marine organisms which have fallen and accumulated within the muds or ooze, of the deep ocean floors. In time, this is covered by more marine sediments of muds and sands, and the organic debris goes through complex changes under anaerobic conditions to become oil. If the ocean floor is uplifted to become land or shallow sea basins, the oil, along with gas and trapped water will migrate upwards from its source rock until it reaches the surface forming tar pits, or becomes trapped by an impermeable cap rock in a variety of geological structures collectively known as **oil traps.**

There are variety of geological structures which can trap oil, gas and water, including anticlinal or dome traps.

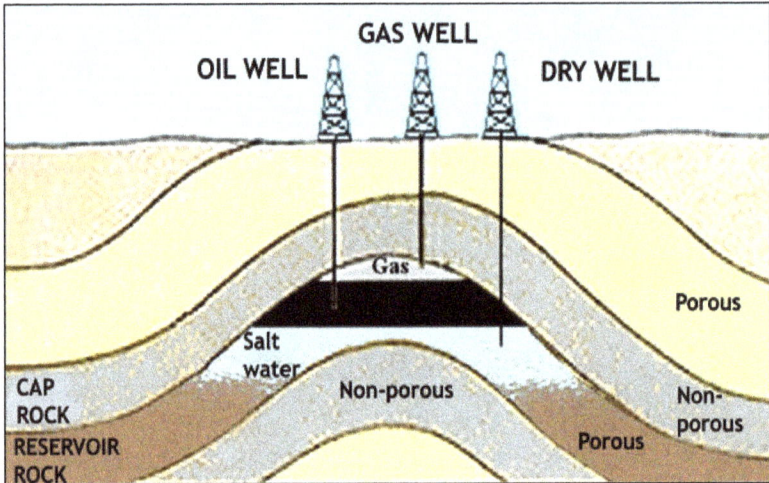

Figure 3.52: An anticlinal oil trap – more accurately it is a 3-dimensional dome.

Oil is usually brought to the surface by drilling down through the rock using oil rigs. These are essentially the same on land, on oil platforms at sea or on drilling ships – a tall derrick is constructed so that sections of pipes or **sticks** are hauled up and attached together using screw-threaded ends. These are then screwed onto the drill bit through a polygonal slot or **kelly** which is set in a rotary table on the raised floor of the derrick. This table is driven by a motor and drilling mud is also pumped down the drill pipe, or casing, to act as a lubricant and coolant. As the bit drills down into the rock, more sticks are added by screwing them into the thread at the top of the previous pipe. When oil or gas is reached, it is usually under pressure so the drill pipe is then attached to a series of pipes and taps called a **christmas tree**. As the pressure reduces, the oil (or gas) must then be pumped out.

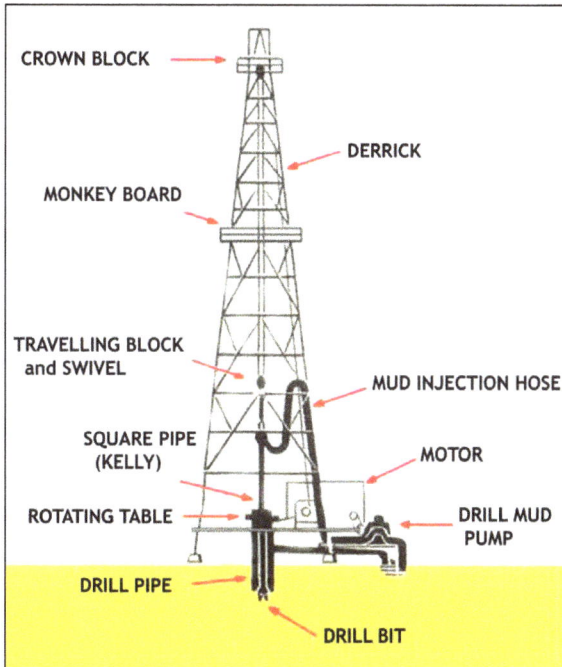

Figure 3.53: An oil rig showing the main parts

Natural gas from petroleum gas wells also contains methane along with heavier hydrocarbons such as ethane C_2H_6, propane C_3H_8, butane C_4H_{10}, pentane C_5H_{12}, higher carbon hydrocarbons, hydrogen sulfide (H_2S) and carbon dioxide which must be expensively separated or removed. Methane gas has always been an explosion hazard in underground coal mines and coal mine methane CMC, once called firedamp from the German: *dampf* for vapour, is usually extracted from mines by large ventilation systems during mining. Methane is a product of the breakdown of the organic matter within coal and is stored within the coal matrix as a liquid adhering to the walls of the pore spaces, a process called **adsorption**. The open fractures in the coal or **cleats**, can also contain free gas or can be saturated with water.

Potentially commercial **coal seam gas (CSG),** containing mostly methane gas, is usually found in seams saturated with water. Coal seam gas contains about 95-97% methane gas (CH_4) and a little carbon dioxide, ethane and nitrogen. The CSG is extracted by means of wells that are drilled down into and along the coal seams of depths of up to 1000 metres. When water is pumped out of the coal seams the confining pressure is reduced, leading to the evaporation of the gas so that it can be collected and gas compressors are usually used to push released gas to a central gas processing facility where it is dried and compressed for transportation along a high pressure pipeline to shipping points.

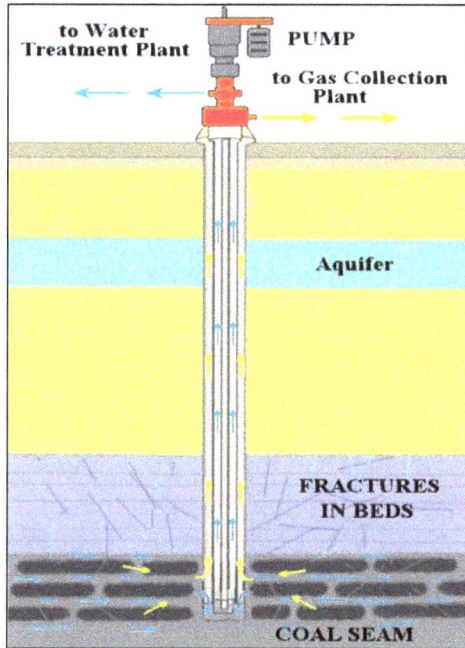

Figure 3.54: A coal Seam Gas well

3.7 Sedimentary Structures

A number of features or structures can often be observed within sedimentary rocks. They may be formed within the loose sediment which has been recently deposited and often indicate the environment of deposition. These can be called **primary sedimentary structures** as they form before the sediments have become rock. Some sedimentary structures may be formed after or near the end of lithification and are called **secondary sedimentary structures**, although there is often some conjecture about when these have been formed.

- Primary sedimentary structures are usually formed as part of the sedimentation process and include:

  - Stratification, or the layering of the sediments as they are deposited, is an identifying feature of most sedimentary rocks. This can be found as a variety of shapes of different sizes depending upon where and how the loose sediment was deposited. A bed is the basic unit of strata which may contain smaller bands and eye-shaped lenses. Beds may be massive with no major internal features whilst others may be inter-bedded. These may have small bands representing minor changes in sedimentation, such as a band of conglomerate that may represent a sudden increase in water flow. Some rocks, such as shales may also fine internal bands or laminae and others simply show the bedding planes which represent the start and end of the bed's deposition.

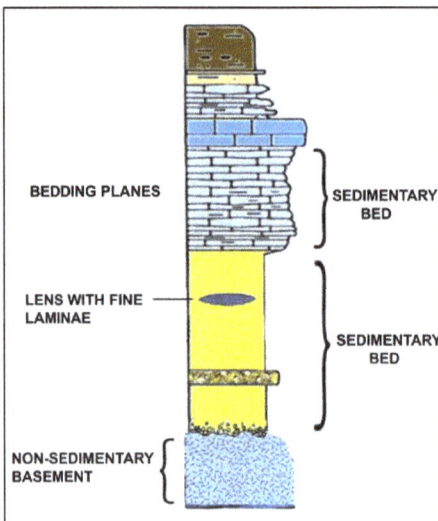

Figure 3.55: A stratigraphic column showing sedimentary bedding

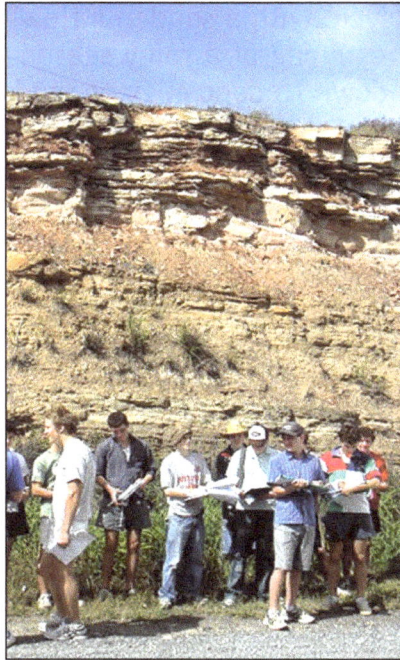

Figure 3.56: Students studying well-bedded strata of sandstones and shales, Ipswich, Australia.

- Graded bedding is seen in many cross-sections. Strata may show grading. This means that in one bed, the sediments at the bottom are coarser than those above. For example, there may be a layer of large-clast conglomerate fining upwards through coarse sand, fine sand and perhaps muds. Although the latter is often absent in a stream with a good flow rate. These graded beds are often found in turbidite deposits and show how in the sudden flow of debris off a coastal shelf, the bigger particles will settle first, followed by finer sediments in order of size. Graded bedding could also represent a regular stream cycle with a sudden inflow of large pebbles at the beginning of a rainy season with the stream slowing down to deposit finer sands as it returns to a more normal rate of flow.

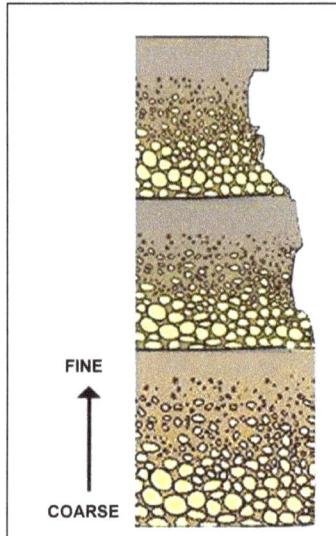

Figure 3.57: A diagram showing three graded beds in a stratigraphic column.

- Cross bedding or **current bedding**, is the pattern of lines and curves called **foresets** seen within a bed of rock, especially sandstones, making an angle to the bedding planes. It represents the piling up of sediment by the ancient flow (palaeocurrent) of water or air which deposited the clasts. It is an important feature to be found, for it gives an indication of speed and direction of the current of the medium of transport either as wind or water. For an accurate estimation of these factors, the geologist must find a true cross-section of the bed showing this feature.

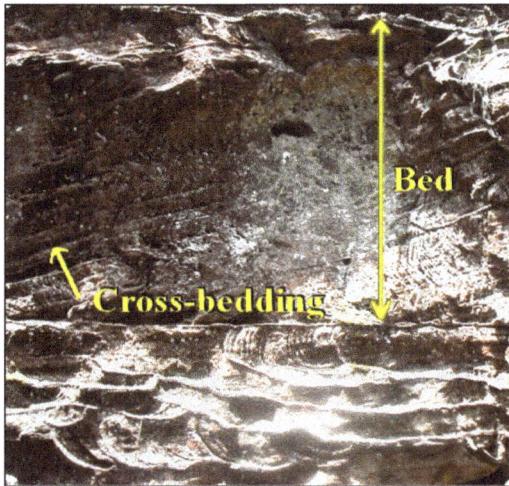

Figure 3.58: Cross-bedding in sandstone, Ipswich Coal Measures, Kholo, Queensland, Australia.

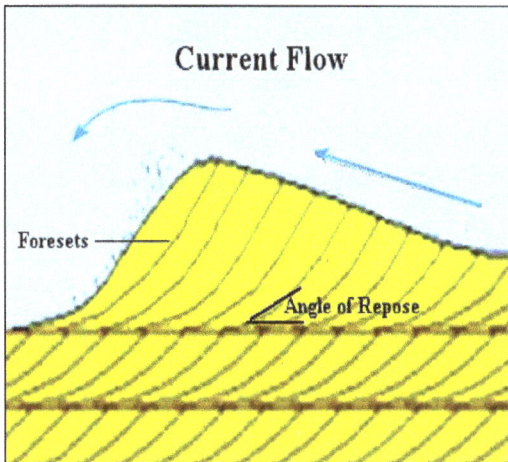

Figure 3.59: Diagram showing how cross-bedding may form water or wind currents

The angle that the foresets make to the horizontal bedding plane is called the **angle of repose**, and represents the steep side on the lee (away from the wind) side of the structure. In aeolian (wind) deposited sandstones, this angle can be over 30° but it is usually a lot smaller for fluvial (stream) deposition.

- Ripple marks – are formed on the surface of a layer of sediment by the action of wind or water currents. These may be preserved on the top of a bedding plane if the ripples are quickly covered with more sediment. There are two main types of ripple marks:

  - Symmetrical ripple marks which have rounded crests and roughly equal slopes on each side indicating two-way water stream such as the ebb and flow of waves or tides.

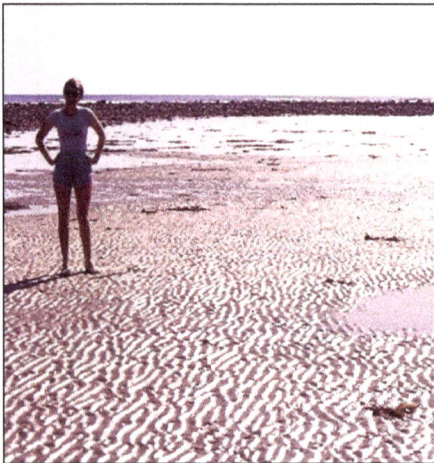

Figure 3.60: Symmetrical Ripple from water action Broome's Head New South Wales, Australia

- Asymmetrical ripple marks show one side of the crest steeper than the other, suggesting one direction of current by wind or water. The shape of the ripple (its curvature and long side) can indicate the direction of the current which made it.

Figure 3.61: Asymmetrical ripples on the top of a bedding plane of mudstone with the stream flowing towards the top of the page, near the Morado Glacier in the Andes southest of Santiago, Chile

- Mud cracks occur when any fine-grained sediment is exposed to the air and dries out. As it dries, the mud shrinks in towards common centres and so the surface breaks up into a series of polygonal shapes. On the top of a bedding plane, mud cracks indicate a period when the pond or lake has dried up. This is very common in small lakes in fluvial conditions but can be extensive in the playas of deserts.

Figure 3.62 Mud cracks in the desert near Uyani, western Bolivia
(Photo: Matthew Scott)

- Tool marks and trails are indentations made in the freshly-deposited sediment before it is covered and hardened into rock. Tool marks are made by pieces of rock, timber or any obstruction that may be moved or swirled around by the current. Relic pot holes may be found where there has been turbulence in the stream or marine rock platform and scouring on the leeward side of isolated ventifacts (wind sculptured rocks), can be found in old desert sands. Animal tracks and tubes left by burrowing animals such as worms or trilobites are sometimes found in mudstones which were once soft sediment on the sea floor.

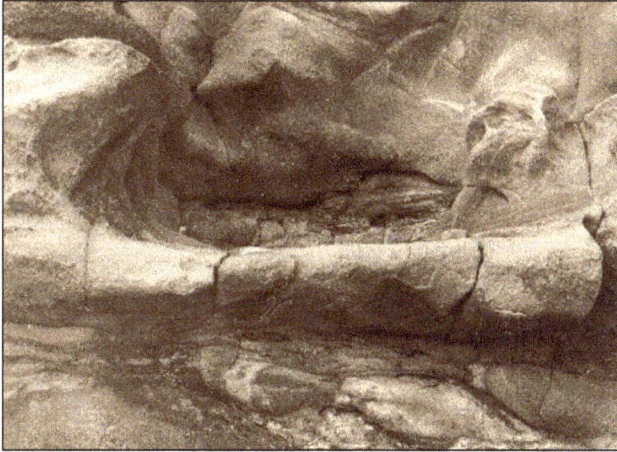

Figure 3.63: Relic pothole (about 30 cm wide) in sandstone bedding, Evans Head, New South Wales, Australia.

- Secondary sedimentary structures are formed after the sedimentary beds have been lithified. They include:

  - Ironstone concretions and bands which occur as small to large rounded balls of a deep rusty red colour due to their composition of iron oxides (mainly of haematite). They may be hollow, and sometimes contain a smaller iron oxide ball. There is some doubt as to how these may be formed but as they are often found near iron-stained sandstones in dry climates, they may be a result of water percolation, the concentration of iron solutions and then their oxidation near the surface which is then eroded to expose the concretions.

Figure 3.64: A hollow ironstone concretion

The sequence of ironstone concretion formation may be as follows:

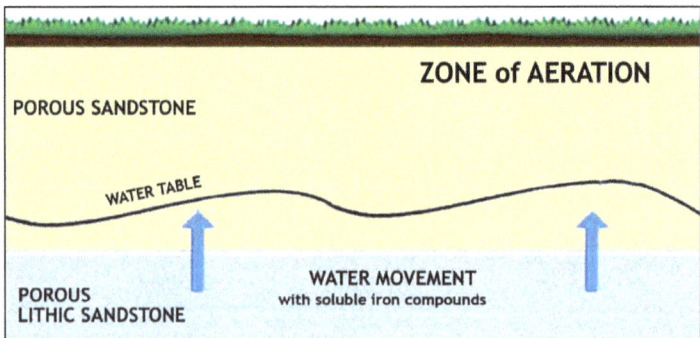

Figure 3.65: Water movement brings up soluble iron compounds which have been leached out of the rock clasts in the lithic sandstone.

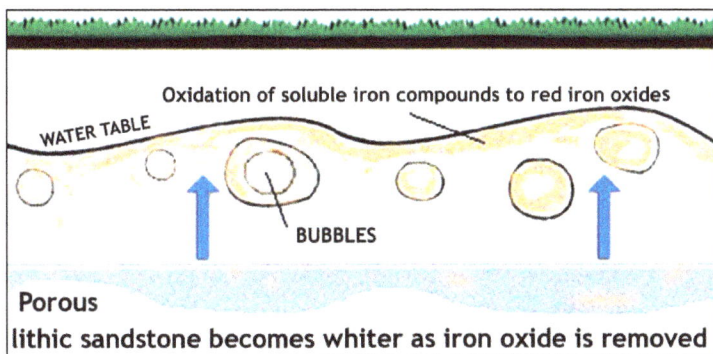

Oxidation of soluble iron compounds to red iron oxides

WATER TABLE

BUBBLES

Porous
lithic sandstone becomes whiter as iron oxide is removed

Figure 3.66: The rising water containing these solutions may not be uniform and consisting of bubbles of liquid much like water in a sponge. The iron solutions in these bubbles oxidise as they come into air contact.

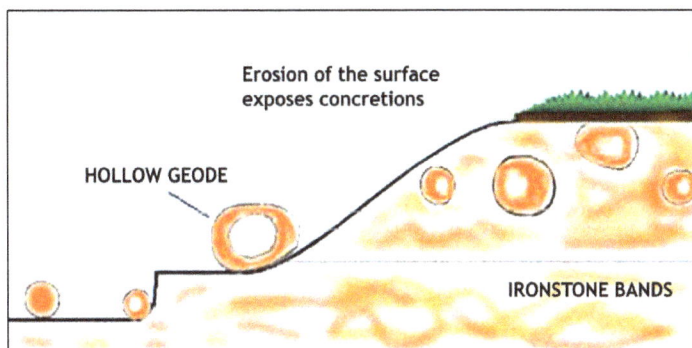

Erosion of the surface
exposes concretions

HOLLOW GEODE

IRONSTONE BANDS

Figure 3.67: Further oxidation and erosion exposes these rounded fossil bubbles as rounded, hollow concretions.

- The leached sandstone may also show some bands of ironstone as a stain which from a distance, sometimes resembles folded or tilted strata.

Figure 3.68: A small block of cut sandstone showing ironstone staining resembling small folds.

- Halite casts are formed in marine conditions when muds or sands are exposed to the air such as on a tidal flat and seawater, saturated with salt, this crystallizes out as cubic crystals of salt (halite). This is then covered over rapidly by new sediment and dissolved away with sediment filling the spaces left.

Figure 3.69: A small specimen of marine sandstone showing halite casts

- Zebra rock is a reddish-brown siltstone with distinctly coloured, intermittent bands and stripes of darker red-purple. The only place in the world where this stone has been found is the Lake Argyle catchment area of the Kimberley Range in Northern Territory, Australia. It is composed of small particles of the minerals quartz and sericite mica of a fine-grained white colour, as well as clay minerals. The colour banding is probably formed by the rhythmic precipitation of iron oxide as haematite, in rich bands during the alteration of the rock by percolating fluids. This stone is often cut and polished as a gemstone.

Figure 3.70: Small blocks of cut Zebra Rock

Chapter 4: Metamorphic Rocks

4.1 Introduction

This is the third major type of rock which makes up the Earth's lithosphere. They are rocks which have been formed from existing rocks which have been changed by heat, pressure or both whilst in the solid state. The name metamorphic comes from the Greek *meta* for change and morphe for form and previously existing rocks, or **protoliths**, such as those of igneous, sedimentary and even metamorphic origin, may suffer change to form entirely new rocks.

BASALT   AMPHIBOLITE

Figure 4.1: The volcanic rock, basalt is metamorphosed to amphibolite

Figure 4.2: The deep intrusive igneous rock granite is metamorphosed to gneiss

Figure 4.3: The sedimentary rock, quartz sandstone metamorphosed to quartzite (both x 1/4)

LIMESTONE → MARBLE

Figure 4.4: When the monomineralic sedimentary rock limestone is metamorphosed to marble, the basic composition of calcium carbonate does not change (both x 1/4)

Some of the common metamorphic rocks formed from their protoliths are given in the following table:

PROTOLITH (ORIGINAL ROCK)	LOW TEMPERATURES & PRESSURES (>200^{0C}, 1.5KILOBARS)	MEDIUM TEMPERATURES & PRESSURES	HIGH TEMPERATURES AND PRESSURES	EXTREME TEMPERATURES AND PRESSURES (> 700^{0C})
SHALE	SLATE	PHYLLITE SCHIST GNEISS		
SANDSTONE	compaction	QUARTZITE		Complete Melting
LIMESTONE		MARBLE		
PERIDOTITE	SERPENTINE	No further change		
BASALT	No change	AMPHIBOLITE, some SCHISTS		
GRANITE	No change		GNEISS	
BLACK COAL	ANTHRACITE		GRAPHITE	

Table 4.1: Simplified chart showing the formation of some of the major metamorphic rocks.
Note: 1 bar of is equal approximately to 1 atmosphere of pressure, 1kilobar (kb) = pressure at 3km deep

4.2 Classification of Metamorphic Rocks

Metamorphic rocks can be grouped or classified by different features or characteristics. These include:

- Texture, and in particular whether the rock is **foliated**, that is, has a flaky appearance due to crystals such as micas being flattened and aligned, or **non-foliated** where there does not appear to be any orientation of crystals which are more granular than flaky. The latter texture usually gives a uniform appearance which may be very fine-grained and smooth to **saccharoidal** i.e. grainy like coarse sugar.

- Type of the metamorphic process which caused the changes in the parent rock, such as contact with heat from igneous bodies or over regional areas through intense pressure and often also heat. Many, but not all of the regional metamorphic rocks, are foliated and most of the contact metamorphic rocks are non-foliated.

- **Facies** or the assemblage or common association of specific minerals which are formed at known temperatures and pressures. These are therefore good index minerals indicating the precise conditions of temperature and pressure.

4.3 Changes in Rock Due to Metamorphism

During metamorphism, there may be changes in:

- Texture is the overall appearance of the rock produced by the size, shape and arrangement of grains or crystals.

- Hardness is the ability of the rock to resist indentation or breakage. Whilst there may be some relationship between the hardness of minerals within the rock this should not be confused with Mohs' Scale of Hardness for minerals. Usually metamorphic action makes the original rock harder and more resistant, although some pressure influences may allow a preference for flat minerals which would give the new rock a flaky appearance and make it more friable.

- Porosity relates to the number of small open spaces within the rock. Metamorphism usually compacts and closes many of these spaces. Permeability, or the ability of a rock to allow fluids to pass between pore spaces, is also reduced.

- Mineralogy of the original rock due to metamorphism by recrystallization. During the metamorphic process, original minerals may be unstable in the new environment and react with the surrounding material to form new minerals which are stable for that changed environment of heat and pressure. Even if conditions revert to those of lower temperatures and pressures, some metamorphic minerals, which are said to be **metastable** (able to exist at those conditions), may remain.

4.4 Controls of Metamorphism

The main factors which control metamorphism in rocks are:

- Temperature which appears to be the critical factor and a temperature of only about 200°C to 300°C is enough to start metamorphism. Over 700°C, however re-melting may take place giving high temperature metamorphics called **migmatites**. Eventually with higher temperature, the rocks melt forming magma and the rock cycle begins again. Increases in thermal energy can be caused by:

  - Geothermal activity or general heat from deep within the Earth increases about 25°C for every kilometre of depth. The **geothermal gradient**, is the increase in heat per unit depth into the Earth and varies with the shape of the heating body. This also produces a curved line of demarcation or **geotherm**. Where there is high surface heat flow, such as areas of active volcanism or mantle plumes beneath thin continental crust, geothermal gradients of 40°C to 100°C per kilometre occur, allowing for relatively high temperatures at relatively shallow levels of the crust. Within the interiors of the thicker continents, geothermal gradients of 25°C to 35°C per kilometre are more common, and at subduction zones where cold oceanic crust is suddenly pushed down to great depths, geothermal gradients range from 10°C to 20°C per kilometre. These variations in heat produce a variety of different metamorphic controls and rock types.

- Radioactive decay naturally occurs in many rocks which contain radioactive isotopes. This is especially found in deep crystalline rocks such as old granites where there may be more concentrations of the heavier radioactive isotopes such as uranium.

- Igneous intrusions of molten magma come up from below in structures such as batholiths, stocks, dikes and sills. These make contact with the country rock through which they pass and cause changes at the margins depending upon the size of the intrusion and its temperature.

- Friction during faulting or plate interaction, especially at convergent margins occur where one plate may move over or below another. It also may occur along conservative margins where plates grind past one another. Here, and along smaller fault lines, the friction effects not only grind rock material, it may also provide enough temperature to cause metamorphism.

- Pressure is also a modifying factor, which may reduce the critical temperature needed to cause metamorphism if it is high enough. The weight of rocks above or the load pressure, or the stress of **tectonic** events such as large earth movements, especially at where plates collide, may provide the pressure necessary to cause metamorphism. Then there is re-packing of the crystal lattice due to basic compression and this causes orientation of flat minerals such as micas and chlorites. Because of their flat crystals, these minerals are able to resist pressure and become stacked at right-angles to it. This gives the rock its typical foliated texture. For most crustal rocks, the weight of the rocks above would produce a pressure approximately equal to one kilobar (1 **bar** = 100 kilopascals of pressure) at a depth of approximately 3.5 kilometres. Most continental crusts are about 30-40 kilometres thick. However, under mountain belts such as the Alps, Andes and Himalayas, the thickness of crust is about 60-80 kilometres and the load pressure is very much greater at the base of the crust, being about 10-20 kilobars. Oceanic crust is generally only about 6-10 kilometres in thickness, and metamorphic pressures are therefore considerably less than in continental regions. In subduction zones, where oceanic and continental crust may be pushed down to depths of over 100 kilometres, metamorphism at very high pressures may occur. In addition, a change in pressure may cause some chemical reactions between existing minerals and/or with those in the environment to produce a new composition.

Online Video 4.1: Thin-section (x polars) of micas in schist
Go to https://youtu.be/C96aLAYDQF4

- Composition or the chemistry of the original rock will also control the nature of the metamorphic minerals which can form under various conditions of temperature and pressure, by providing the raw material of elements which can move (as ions) and then recrystallize as new minerals. Some metamorphic rocks, such as quartzite and marble contain only the same mineral as in their parent rock (quartz sandstone and limestone respectively). Such a change is called **isochemical metamorphism** (Greek: *iso* – same).

- Movement of hot fluids, especially hot sub-surface water and carbon dioxide dissolved in it. These solutions form as minerals in the original rock break down as they are heated with the expulsion of water, carbon dioxide and other dissolved ions. These may then penetrate into the surrounding rock causing changes or **metasomatism**. This control is probably only a relatively minor one, but it does contribute to the suite of metamorphic rocks formed by the hydrothermal alteration of existing rocks by hot solutions. For example, skarns are calcium-silicate rocks formed by hot solutions entering limestones, dolomites and other carbonate-rich rocks from nearby granite intrusion.

4.5 More on Texture

Texture refers to the way that different minerals are formed and orientated within the rock. Because the minerals in metamorphic rocks form within a solid medium, i.e. there is no re-melting of the minerals, only the physically stronger minerals such as olivines, andalusite, sillimanite and garnets will form well-shaped crystals with regular faces called **idiomorphic crystals**. Other, weaker minerals, especially feldspars and micas will form as flakes or grains with no regular crystal faces. These are called **xenoblastic crystals**, but they may give the rock some foliation. Textures generally reflect the nature of the metamorphism and the most common metamorphic rock textures include:

- **Hornfelsic** is a non-foliated texture, usually fine grained, with crystals uniform in size and shape, and randomly oriented e.g. hornfels and skarns.

Figure 4.5:   Hornfels specimen showing the typical sheen of the micas in hornfelsic texture

- **Granular** or granoblastic texture is a non-foliated texture with well-formed crystals of equal size with well-defined edges which may often be **sutured** i.e. with crinkled edges looking like they had been stitched together. Quartzites and granulites are high temperature rocks from deep sources and contain coarse grains of quartz, feldspars and ferromagnesian minerals. If the grains are very coarse, they may be referred to as **saccharoidal** (like sugar) e.g. quartzite, marble and granulite.

Figure 4.6: Quartzite in hand specimen

Figure 4.7: Quartzite in thin-section (cross polars x 25) showing interlocked grains of quartz

Figure 4.8: Saccharoidal marble showing coarse crystals of calcite

- **Slaty** is a foliated texture of very fine grains, often too small to see with the naked eye, with orientation of crystals to give flat layers which are easy to split apart. This texture is the first in the order of increasing pressure of the foliated metamorphic rocks e.g. slates.

Figure 4.9: Slate specimen showing the flat layers of slaty texture

- **Phyllitic** is a foliated texture with fine grains just visible and flattened together. Mica crystals give these rocks a slight sheen e.g. phyllites.

Figure 4.10: Phyllite specimen showing the slight sheen of new mica crystals

- **Schistose** is a foliated texture with large, flaky crystals of micas, chlorites and sericite minerals giving a great variety of schists named after the main mineral e.g. biotite schist, chlorite schist, sericite schist.

Figure 4.11: Chlorite schist with the shiny green foliation of chlorite mineral

Figure 4.12: Biotite schist with large flakes of biotite mica

- **Gneissic** is a coarsely foliated texture with very large, coarse-grained flattened crystals of micas, feldspars and quartz. Often extreme movement of ions during metamorphism may lead to concentrations of minerals or **porphyroblasts**, within the rock, giving it a spotted appearance. Examples of porphyroblastic textures are biotite "eyes" or **augen** in augen gneiss, cordierite porphyroblasts in spotted hornfels and garnets in schists and gneisses.

Figure 4.13: A typical gneiss

Figure 4.14: A garnet gneiss showing pink garnets

4.6 Types of Metamorphism

There are several types of metamorphism depending upon the processes of change due to the original mineralogy of the parent rock and the conditions of temperature and pressure which are applied. The most common types of metamorphism are shown in the following graph:

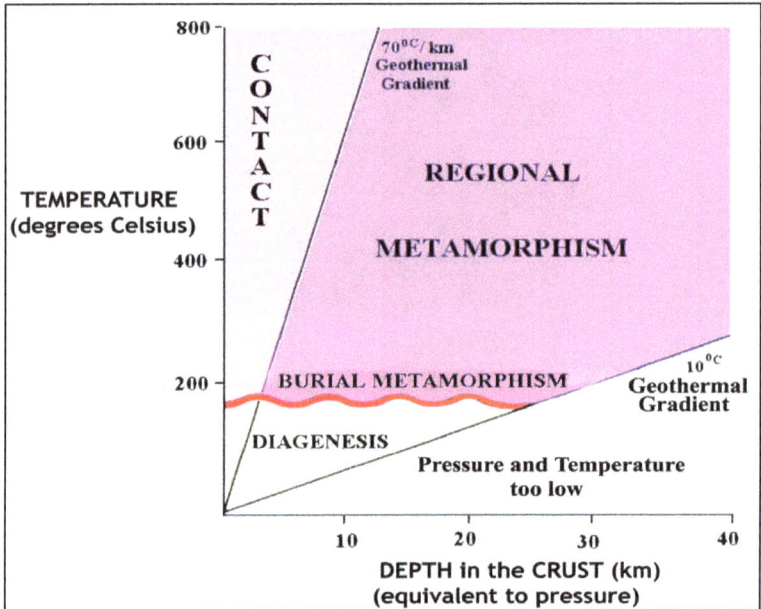

Figure 4.15: A graphical representation of types of metamorphism

The most common types of metamorphism include:

- **Burial metamorphism** occurs where sediments are buried at depth with little, if any, folding and intrusion of igneous bodies. The **lithostatic pressure** of the weight of the rock material above further compresses the rock below. Textures of the rock remain the same as the original, but some new minerals, such as zeolites and illite, a mica-like mineral formed from clays, may form. Rocks become less porous, more compact and

146

some water is removed. Burial metamorphism represents the lowest grade of metamorphism and is an extension of the lithification of sediments or **diagenesis**, within basins of deposition, but at increased pressure, usually greater than 3 kilobars, and temperatures greater than 200°C.

- **Contact metamorphism** is due to changes due to heat from contact with some igneous body such as a dyke, sill or batholith, which has intruded the parent rock. Very little pressure is involved, and the effect of the metamorphism increases with proximity to the heat source, the size of the igneous body and its temperature. The zone of metamorphism surrounding an igneous intrusion is called a metamorphic **aureole**, and its size and shape depends upon the size and heat of the intruding body. Aureoles can vary in size from a few millimetres near thin dykes, to many kilometres around very large batholiths.

Apart from the effects of the high temperature of the intrusion, hydrothermal solutions and gases may move from the intrusion into the parent rock to produce hydrothermal alteration.

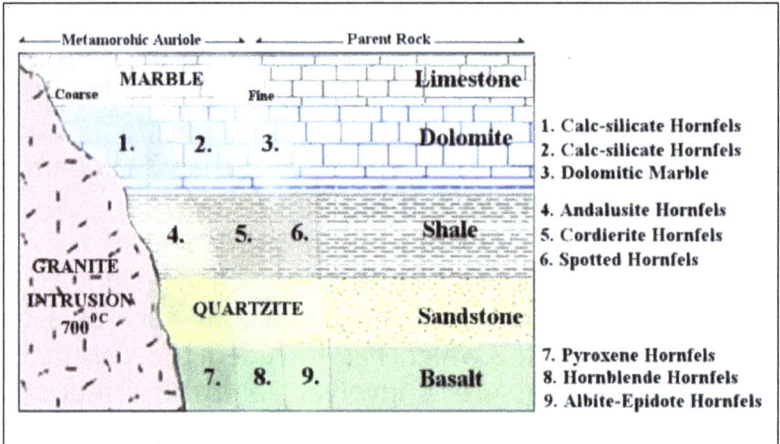

Figure 4.16: Diagram showing the zones of metamorphic rocks formed from parent rocks by an igneous intrusion of granite

In addition, chemical ions and gases may also be removed from the parent rocks to give a wide range of new minerals within the contact aureole. There also may be veins of quartz and feldspar in the rocks surrounding the igneous body which contain valuable economic minerals such as gold, silver, chalcopyrite, molybdenite and galena. Since there will be a gradual increase in metamorphic change in minerals approaching the intrusion, and that these changes will also depend upon the original mineral composition of the parent rock. Zones have been defined to indicate the degree of metamorphism. These zones can be defined in terms of various mineral facies consisting of a specific set of characteristic minerals which come from parent rocks of mafic composition e.g. the basalt in the example given in the diagram above. These zones can be defined as the:

- **Albite-epidote hornfels facies - lowest grade of metamorphism**
- **Hornblende hornfels facies**
- **Pyroxene hornfels facies - highest grade**

Each facies has its characteristic set of minerals or **index minerals**, which appear at the specific temperatures for each type of parent rock:

FEATURES	FACIES		
	ALBITE-EPIDOTE HORNFELS	HORNBLENDE HORNFELS	PYROXENE HORNFELS
Textures	Original + spots	Coarse + increasing Porphyroblasts	⟶
Structures	Original becoming slaty, schistose	increasing Flinty	⟶
Temperatures (degrees Celsius)	300	500	700

Composition	Index Minerals			
Pelitic Rocks Rich in **Aluminium** and Silica and Clays e.g. SHALES	ALBITE EPIDOTE QUARTZ ANDALUSITE MUSCOVITE BIOTITE ORTHOCLASE PLAGIOCLASE SILLIMANITE			
Calcareous Rocks with Carbonates and impurities e.g. Limestones & Dolomites	TALC TREMOLITE QUARTZ CALCITE DIOPSIDE GROSSULAR WOLLASTONITE PLAGIOCLASE			
MAFIC Rocks with Iron and Magnesium minerals e.g. Basalts	QUARTZ ALBITE EPIDOTE CHLORITE HORNBLENDE PLAGIOCLASE ACTINOLITE DIOPSIDE HYPERSTHENE			

Table 4.2: Showing the index minerals for different parent rock types and metamorphic facies – the lengths of the lines (at right) indicate the stability of the index mineral within the rock e.g. when limestones are altered by contact metamorphism, the mineral plagioclase will indicate a temperature 300^{OC}.

There is also another facies which is rare and beyond the usual range of the highest temperatures normally encountered in contact metamorphism – this is the **sandine facies**. This facies is at the edge of complete re-melting of the rock to form magma and thus begin the rock cycle anew. It consists of highly altered rock with some molten sections which quickly revert to non-minerallic glass. Minerals in this facies include cordierite, mullite, sanidine and the mineral tridymite which is often altered to quartz.

As the temperature rises, unstable minerals react with their surroundings to form new minerals which are stable at the new temperature. For example:

For impure limestones:

$$Ca\ CO_3 \quad + \quad SiO_2 \quad = \quad Ca\ SiO_3 \quad + \quad CO_2$$

calcite   quartz   wollastonite   carbon 500^{oC}
dioxide gas

For dolomites:

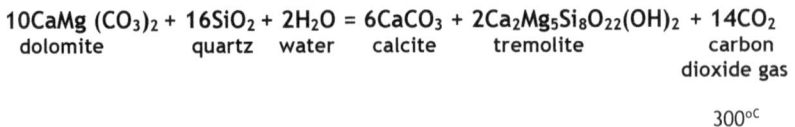

$$10CaMg\ (CO_3)_2 + 16SiO_2 + 2H_2O = 6CaCO_3 + 2Ca_2Mg_5Si_8O_{22}(OH)_2 + 14CO_2$$

dolomite   quartz   water   calcite   tremolite   carbon
dioxide gas

300^{oC}

For monomineralic parent rocks, that is those rocks having only one predominant mineral, such as quartz in sandstone and calcite in limestone, contact metamorphism only produces changes in texture and other physical properties such as porosity, hardness, density, rather than producing any new minerals.

- **Regional metamorphism** (or Dynamic Metamorphism) involves a series of complex physical and chemical changes which occur when large rock masses are subjected to pressure, usually with high temperatures over a very large area. Such metamorphism is produced in orogenic belts at convergent plate margins where intense folding, or subsidence to great depth or both, produces the characteristic changes. Pressures are normally above 2000 bars with temperatures above 200°C. Generally, rocks of regionally metamorphic belts have foliated textures typical of strong stresses and strong deformation. Schists and gneisses are the most common rocks of such areas and several grades or zones of different metamorphic rocks may be produced from the one parent rock.

Figure 4.17: Increasing grades of metamorphism

In 1912, the eminent geologist **George Barrow** (English: 1853-1932), noted several zones of minerals in the metamorphic rocks of the Scottish Highlands. He set down six zones of classification which could be used in regionally metamorphic rocks of other countries. These zones were predominantly for the metamorphism of shales and similar rocks, and each zone had its characteristic identifying mineral which appeared first during the metamorphism of the preceding rock. His zones were the:

chlorite zone (lowest metamorphism)
biotite zone
garnet zone
staurolite zone
kyanite zone
sillimanite zone (highest metamorphism)

A modern system of classification is that of regional metamorphic facies. This is mainly defined on temperature and considers groups of minerals which are characteristic of various grades of metamorphism. As research continues on the complexity of mineralization and controls in metamorphism, metamorphic petrologists have developed several models of metamorphic facies classification. The main facies for regional metamorphism are given in the table below:

FACIES	INDEX MINERALS IN MAFIC ROCKS
ZEOLITE FACIES Lowest temperatures & pressures of regional metamorphism	heulandite, analcite, quartz, (± clay minerals) laumontite, albite, quartz (± chlorite)
PREHNITE-PUMPELLYITE FACIES Slightly higher temperatures & pressures	prehnite, albite, quartz; pumpellyite, chlorite, epidote, albite, quartz; pumpellyite, epidote, stilpnomelane, muscovite, albite quartz
GREENSCHIST FACIES Medium temperatures & pressures. Schistose texture and green in colour	chlorite, albite, epidote (± actinolite, quartz) albite, quartz, epidote, muscovite, (± stilpnomelane)
AMPHIBOLITE FACIES Medium pressure and average to high temperatures	hornblende, plagioclase,(± epidote), garnet, cummingtonite, diopside, biotite
GRANULITE FACIES Highest grade of metamorphism at medium pressure	orthopyroxene, clinopyroxene, hornblende, plagioclase (± biotite); orthopyroxene, clinopyroxene, plagioclase (± quartz); clinopyroxene, plagioclase,garnet, ±orthopyroxene (higher pressure)
BLUESCHIST FACIES Low temperature but high pressure, such as occurs in rocks in subduction zones. Glaucophane is blue with schistose texture.	glaucophane, lawsonite, chlorite, sphene, (± epidote ± phengite ± paragonite), omphacite
ECLOGITE FACIES Highest pressures and highest temperatures of regional metamorphism	omphacite, garnet, (± kyanite), quartz, hornblende, zoisite

Table 4.3: Major regional metamorphic facies and their index minerals

These facies are controlled by temperature and pressure and indicate precise modes of formation.

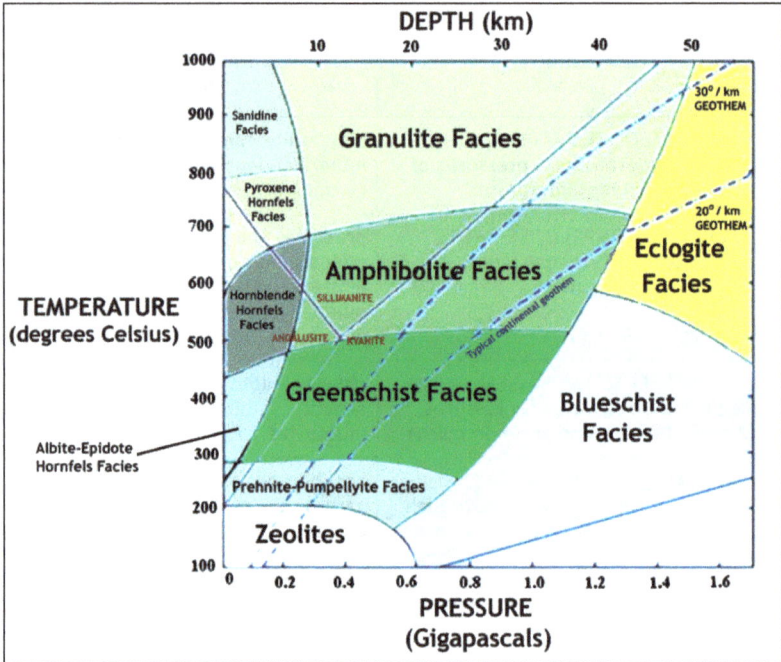

Figure 4.18: A phase diagram (graph of the different forms of substances at different conditions) showing the main controls and facies for both contact and regional metamorphism

A good example of how minerals change during metamorphism is shown in the graph (above) as a **triple point** between the minerals kyanite, andalusite and sillimanite. This is an example of a one-component system, as all three minerals are **polymorphs** of aluminum silicate (Al_2SiO_5), that is, they have the same chemical composition but different crystal lattices.

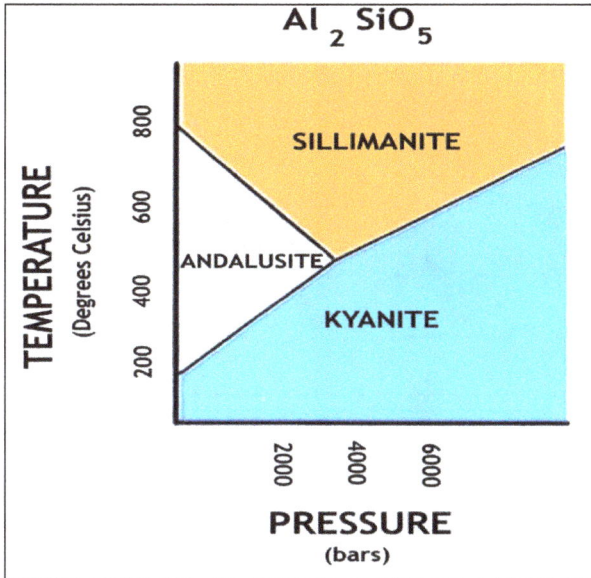

Figure 4.19: Closer view of the phase diagram showing the
triple point for the polymorphs of Al₂SiO₅

The phase diagram above (Figure 4.19), shows that each
of the minerals is stable over a particular range of
temperatures. Andalusite, only forms at pressures below
the triple point (the unique point where the three minerals
co-exist at approximately 4000 bars and at about 500^{0C})
and sillimanite only forms at temperatures above the
triple point. At pressures less than that of the triple point,
a rock composed of $Al_2 SiO_5$ would first produce kyanite,
then andalusite and finally sillimanite. This diagram shows
that mineral assemblages observed in metamorphic rocks
may be used to estimate the pressure and temperature at
which a specific rock has formed.

Figure 4.20: Grey, fibrous sillimanite

Figure 4.21: A brown andalusite crystal embedded in quartz and other minerals

Figure 4.22: Shiny blue kyanite crystals within quartz

As a general example of how the main types of metamorphic rocks are formed are those produced at a subduction zone where parent rocks of one plate are pushed below another plate. Here, the parent rocks undergo the various types of metamorphism depending upon the degree of heat and pressure produced as the plate is pushed below the surface.

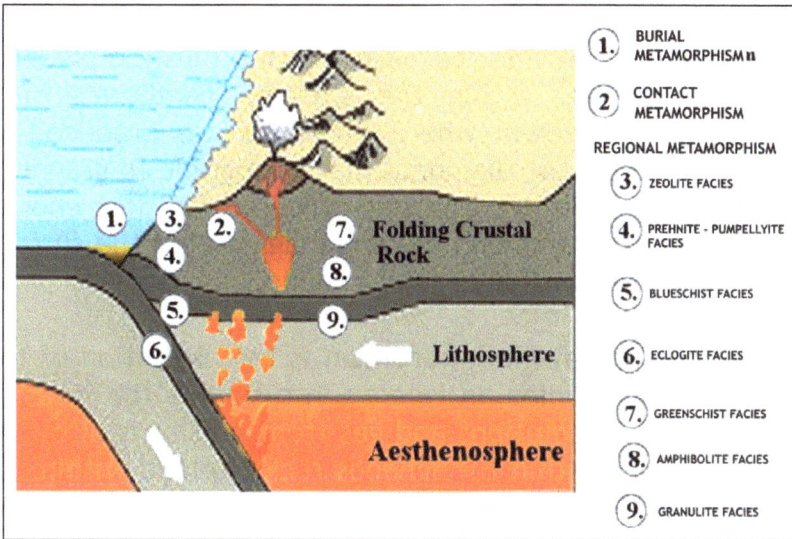

Figure 4.23: Simplified diagram showing the general regions of major metamorphism

4.7 Other Types of Metamorphism

Whilst contact and regional metamorphism constitute the major changes in rocks, there are some other minor forms of metamorphism. These include:

- **Retrograde metamorphism** which is a reversal of the normal prograde metamorphism, and involves the reconstitution of a metamorphic rock to a lower grade of metamorphic stability under decreasing temperatures and usually pressures. This also involves the addition of volatiles such as water and carbon dioxide. This allows the mineral assemblages formed in prograde metamorphism, to revert to those more stable at the less extreme conditions. This is a relatively uncommon process, because volatiles must be present. For example, within the extreme conditions of a carbonite volcano where diamonds are produced at great depth, the diamonds may retrograde to graphite. It is only when the volcanic processes are fast enough to bring the diamond-bearing magma near the surface that these gemstones can be found.

- **Hydrothermal metamorphism** is the alteration of the rocks surrounding an igneous intrusion by the invasion by very hot solutions and gas from it. This is also may also occur within the igneous rocks formed within the intrusion by the hot solutions left at the end of the cooling and crystallization of the rock. This is common in basaltic rocks which generally lack hydrous minerals, those containing water in their crystal lattice. The hydrothermal metamorphism in these rocks result in alteration of mafic minerals to hydrous minerals also rich in magnesium and/or iron, such as talc, chlorite, serpentine, actinolite, tremolite, zeolites, and clay minerals. Rich ore deposits are often formed as a result of hydrothermal metamorphism.

- **Cataclastic metamorphism** occurs as a result of mechanical deformation, such as when two bodies of rock slide past one another along a fault or shear zone. Heat is generated by this friction of sliding action, and the rocks also tend to be mechanically deformed, crushed and pulverized, due to the shearing. Cataclastic metamorphism is not very common and is restricted to a narrow zone along which the shearing occurred.

- **Shock metamorphism** or impact metamorphism are the changes in country rock when an extra-terrestrial body, such as a meteorite or comet impacts with the Earth or if there is a very large volcanic explosion with ultra-high pressures and temperatures being generated in the impacted rock. These ultra-high pressures can produce minerals which are only stable at very high pressure, such as the quartz polymorphs coesite and stishovite. In addition, they can produce textures known as shock lamellae (seen as small layers) in mineral grains, shatter cones in the impacted rock and in extreme cases, tektites, the black, glassy beads formed as the molten rock is thrown up into the atmosphere, quickly cooled and returned to the surface.

4.8 Shield Areas of Continents

Metamorphic rocks are often found in the large areas of ancient rocks called shield areas or zones, found on many continents. These are older Palaeozoic and Precambrian rocks usually much older than about 550 million years. Often referred to as **cratons**, they are characterized by ancient crystalline metamorphic rocks such as gneisses and schists and low relief. Usually, they have an ancient core surrounded by progressively younger rocks due to the **accretion**, or joining and welding together of young orogenic belts onto the older craton by collision of **tectonic plates**. Shield areas are geologically stable, having little new mountain building activity and are usually greatly folded and deformed. They often contain vast mineral wealth due to the concentrations of valuable ores formed by the ancient metamorphism and folding.

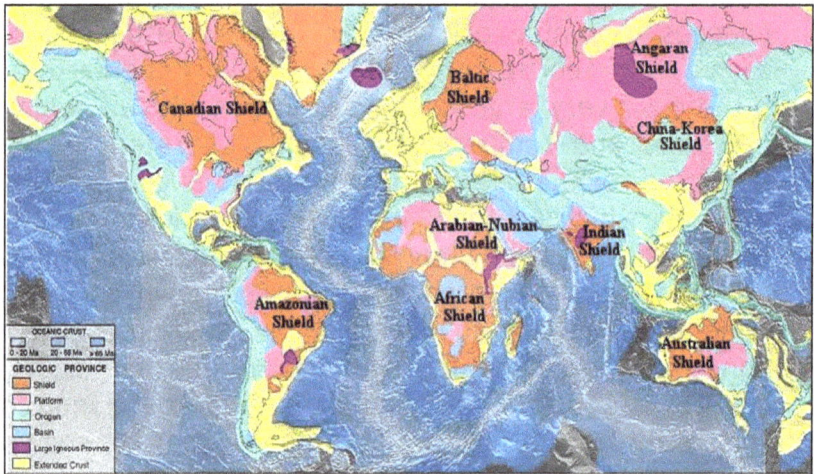

Figure 4.24: Map showing the major continental shields in the world. Not shown is the Antarctic Shield which occupies most of the western part of this continent under 4000 m of ice (Photo: USGS modified)

Summary

1. Igneous rocks are formed from molten rock either deep below the surface from magma (intrusive with large crystals) or on the surface from lava and pyroclastics (extrusive with small or no crystals)

2. Igneous rocks with high quartz and feldspar content are usually lighter in colour and are termed felsic. Dark-coloured igneous rocks have little or no quartz and considerably more iron and magnesium minerals and are termed mafic.

3. Pyroclastic igneous rocks are made from the compaction of broken material erupted out of volcanoes. This material (tephra) can range in size from microscopic dust to large blocks.

4. Volcanoes can give out large volumes of gases from their vents, including steam, carbon dioxide, sulfur dioxide, nitrogen and other gases.

5. When magma cools below the surface it can form intrusive igneous structures such as large batholiths which can cover many hundreds of kilometres when exposed to smaller stocks and bosses.

6. Smaller, tabular intrusions can form dykes (cut across layers of country rock), sills (formed between layers), laccoliths (which form between layers then dome up) and lopoliths (form between layers and then sink in the middle).

7. Volcanoes are extrusive igneous structures because they form on the earth's surface. Volcanoes can be formed from ash (cinder cones), lava (rhyolite domes and shield volcanoes) or a combination of successive ash and lava (stratovolcanoes or composite volcanoes).

8. When igneous rocks cool and solidify, the minerals crystallize within them in specific processes described by Bowen's Reaction Series which shows the order in which minerals form and also which minerals are compatible within certain rocks.

9. Sedimentary rocks are formed by material which has been eroded, transported and then deposited in horizontal layers by water, ice or wind where the velocity of the transporting media has been reduced. They are then compacted by more layers on top and then cemented by minerals present such as silica, calcite, clays or iron oxides.

10. When particles (clasts) settle, they do so in order of size with the heavier particles dropping out first e.g. pebbles, sand, silt, and then clay. They also become more rounded with transportation.

11. In any given sequence of sedimentary rock, the oldest has been deposited first and the youngest last. This is called the Law of Superposition.

12. Sedimentary rocks can be classified as clastic (having visible particles) and non-clastic where no particles are visible and the rock has been formed by chemical or biological processes.

13. Sedimentary rocks which have interconnected pore spaces are permeable and often contain valuable resources of underground water as artesian springs (comes out under pressure) and sub-artesian wells (must be pumped)

14. Oil and natural gas are formed in deep marine conditions from organic remains of plankton and migrate upwards from their source rocks until they are trapped in various structures by impermeable cap rocks.

15. Coal is formed in a freshwater (lacustrine or paludal environments) from the anaerobic breakdown of buried plant material.

16. Sedimentary rocks often show internal sedimentary structures such as cross-bedding, mud cracks and ripple marks which can be used to determine the environment of deposition

17. Sedimentary rocks are valuable as building material (sandstones, limestones) and as sources of water, coal, oil and gas.

18. Metamorphic rocks are formed from existing rocks by the action of heat, pressure or both.

19. Metamorphic rocks form from existing rocks by the influence of heat (usually between 200^{0C} and 700^{0C}) and pressure (up to 20 kilobars with 1 bar being approximately the same as atmospheric pressure). The original rock composition will determine the reactions which produce new minerals and the effects of introduced hot fluids (causing hydrothermal alteration).

20. When the original rock (protolith) is metamorphosed there may be changes in texture (becoming more shiny with micas then with layers for foliated rocks and coarser grained if non-foliated); hardness (generally becoming harder); porosity (less porous); and mineralogy (apart from quartzite and marble which retain their original composition), many new minerals are formed by recrystallization in a solid state.

21. They can be classified by texture (foliated with flaky, semi-layers or non-foliated with smooth or granular appearance), type (contact from hot igneous bodies, or regional by heat and pressure over a large area) or by facies (from the index minerals which first appear).

22. Textures in metamorphic rocks can be: hornfelsic (smooth, small grains e.g. hornfels and skarns), granular (bigger, uniform grains e.g. quartzite, marble and granulite), slaty (flat with hard, smooth surfaces and thin layers which easily split e.g. slates), phyllitic (shine due to small micas e.g. phyllites), schistose (very shiny and flaky due to micas e.g. biotite schists) and gneissic (coarse-grained with flattened crystals e.g. gneisses).

23. Burial metamorphism is low grade and caused by the weight of material above rocks or sediments which have been buried at modest depth (pressure about 3 kb and temperatures just over 200^{0}C) and is an extension of the normal compaction and lithification of sediments (diagenesis).

24. Contact metamorphism is due to heat with minimal pressure and occurs when an igneous body such as a batholith intrudes the country rock. Temperatures are usually between 300^{0C} and 700^{0C} and several facies (specific mineral assemblages defining new conditions) are recognized for the grades of metamorphism.

25. Regional metamorphism occurs over a very large area such as in the mountain building process (orogeny) of continents at convergent plate margins. Here pressures and temperatures can be very high and at the extreme end, melting may occur. The grade of the metamorphism is described by a number of facies having index minerals relating to conditions of temperature and pressure.

26. Other, minor forms of metamorphism, include retrograde (a reversal of the grade due to injection of volatiles and reduction in temperature and pressure), hydrothermal (due to the injection of hot solutions), cataclastic (due to the grinding of rock surfaces together at faults) and shock metamorphism (due to impact from meteors or comets).

27. Many of the world's metamorphic regions occur in the great continental shield or cratons which form the older rocks of the continents.

Practical Tips

1. Igneous landscapes are characterized by sharper-looking, isolated peaks and volcanic areas may have invisible gas vents, hot springs and geysers. Care should be taken in these active areas, especially around lava fields, gas vents and hot springs where the ground surface could be thin and fragile.

2. When in an active area, take note of local advice and do not venture beyond safe limits. Craters and their edges can be very unstable and eruptions very sudden.

3. Old, extinct igneous landscapes are noted for their typical igneous rocks and the red soils derived from them. Often hills may have flat tops due to a capping from an old lava flow and dykes, sills, batholiths and laccoliths may now be exposed to the surface. Old batholiths typically erode to form clusters of rounded boulders (tors).

4. Igneous rocks are usually crystalline (except for the soft, gritty ash and the shiny volcanic glasses) in fresh hand specimen and a good hand lens can be used to identify the rock.

5. Large batholiths can cover hundreds of square kilometres and are seen as groups of eroded tors. The streams around the edges of batholiths, stocks and bosses may contain valuable minerals (gold, silver, gemstones) weathered and eroded from the parent rock.

6. Sedimentary rocks are often very weathered and so it is important to locate good specimen. This can often be done in newly-formed road or rail cuttings but extreme caution is needed in these places.

7. Good specimens should be carefully wrapped in newspaper and inserted into specimen bags and boxes for transportation as they are often soft and friable.

8. In the field, close attention should be given to the size and nature of sedimentary beds from which any specimens may be taken. Often the associated rocks and internal structures will also indicate the environment of deposition.

9. Artesian bores can often be seen at night on extensive inland plains because of the natural gas which also comes up with the water and is ignited by land-owners. Sub-artesian water must be pumped to the surface. When using a tap or faucet connected to a sub-artesian pump, always turn the tap on full. Partial flow in some wells can cause their clay lining to break off and block the shaft.

10. Coal is found within freshwater sedimentary rock sequences so fossil plants, and coaly fragments within grey shales are a possible indicator that coal may be nearby whereas oil is from a marine environment so marine sedimentary rocks, especially sandstones and shales with small microscopic oil drops may be a possible indictor of oil potential.

11. The principle of uniformitarianism – or the present is the key to the past, works well if one is a good observer and can put together several pieces of information read in the rock layers. Structures, such as water or sand ripples, fossils and the nature of the grains within the rock (roundness, sorting and so on) can be used to build up a good picture of the ancient environment.

12. Metamorphic rocks are often found in mountainous areas and the surface may be broken by occasional sharp-edged outcrops so care in exploration is needed. Micas are common in stream beds and in areas where garnet is a major index mineral, these two resist weathering and can be found in stream gravels as flakes and polygonal spheres respectively.

13. Specimens of some schists are very friable (that is they break apart easily) and care is needed in their transportation – wrap and bag very carefully.

14. Many metamorphic rocks such as gneisses, slates, marbles and quartzite are valuable building stones and so are worth finding if in an economically accessible region.

15. Hydrothermal metamorphic minerals and rocks such as skarns often contain valuable economic metallic ores and these metamorphic minerals and alteration features are an aid in location of these ores.

Multichoice Questions

1. An igneous rock typical of lava flows is:

    A. Granite
    B. Basalt
    C. Dolerite
    D. Diorite

2. The type of volcano usually found near plate subduction zones at ocean trenches is:

    A. Basaltic
    B. Rhyolitic
    C. Granitic
    D. Andesitic

3. The igneous rock having little or no quartz, much olivine, pyroxene and calcium pyroxene would most likely be:

    A. Granite
    B. Dolerite
    C. Gabbro
    D. Diorite

4. The following photographs are of different types of volcano. The one which is a shield volcano is:

A.

B.

C.

D.

A. A
B. B
C. C
D. D

5. The following sketch is of a section through a sequence of layers of rock which have been intruded by igneous rocks.

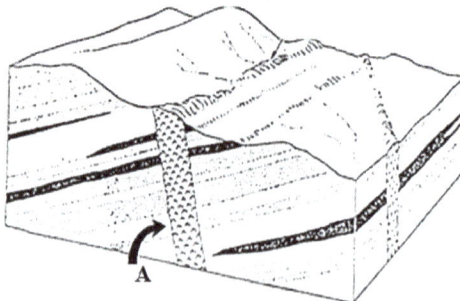

The structure labelled A is most likely a:

A. Sill
B. Laccolith
C. Dyke
D. Batholith

6. The process of formation of sedimentary rocks usually follows the sequence:

A. Erosion – sedimentation – deposition – compaction
B. Erosion – transportation – sedimentation – compaction
C. Erosion – transportation – compaction – deposition
D. Erosion – deposition – compaction – transportation

7. The Law of Superposition refers to:

A. Order of evolution of fossils
B. Overturning of rock strata
C. Rock layers high up on mountain
D. Sequence of formation of strata

8. A rock layer is found to consist mainly of poorly sorted angular fragments, mostly of quartz and rock fragments with small lenses of limestone containing coral fossils. The most likely environment of deposition would be a(n):

A. Offshore trench
B. Beach
C. Desert
D. Glacial valley

9. If a shale is subjected to high temperature (600^{0C}) and high pressure, it would most likely become:

    A. Gneiss
    B. Hornfels
    C. Magma
    D. Schist

10. This question refers to the following facies chart for contact metamorphism:

FEATURES		FACIES		
		ALBITE-EPIDOTE HORNFELS	HORNBLENDE HORNFELS	PYROXENE HORNFELS
Textures		Original – spots	Coarse – Porphyroblasts	increasing →
Structures		Original becoming slaty. schistose	Flinty	increasing →
Temperatures (degrees Celsius)		300	500	700
Composition	Index Minerals			
Calcareous Rocks	TALC	▬▬▬▬		
	TREMOLITE	▬▬▬▬		
with Carbonates	QUARTZ	▬▬▬▬		
and impurities	CALCITE	▬▬▬▬		
e.g. Limestones	DIOPSIDE		▬▬▬▬	
& Dolomites	GROSSULAR		▬▬▬▬	
	WOLLASTONITE		▬▬▬▬	
	PLAGIOCLASE			▬▬▬▬

From this chart, the most likely event would be:

    A. Tremolite would be found in hornfels heated to 500⁰C
    B. Talc and wollastonite can be found in the same rock
    C. Calcite is unstable at low temperatures
    D. Plagioclase and grossular could be found in the same rock

Review and Discussion Questions

1. Distinguish between:

(a) intrusive and extrusive igneous and
(b) mafic and felsic igneous rocks

2. Use Bowen's Reaction Series to explain which other minerals may be found in an igneous rock in association with:

(a)   olivine
(b)   quartz
(c)   anorthite
(d)   biotite
(e)   augite

3. Explain what is meant by the following terms:

(a) batholith
(b) discordant structures
(c) porphyritic texture
(d) accessory minerals
(e) ferromagnesian minerals

4. What are the modes of occurrence (i.e. where they form and how) of:

(a) granite
(b) basalt
(c) dolerite
(d) pegmatite
(e) andesite

5. What would be some indicators that a local area was once a site of igneous activity?

6. Sometimes the source of extrusive igneous rocks is not obvious due to erosion or distance from the field area. What are some ways by which the source of extrusive rocks can be found?

7. Under a microscope, the grains of quartz in sandstone appear to be formed from smaller grains surrounded with overgrowths of quartz (i.e. it looks like a grain within a grain). What is a probable explanation of this?

8. Define each of the following terms:

     (a)   clastic
     (b)   lithification
     (c)   primary structures
     (d)   concretions

9. Indicate the past environments which probably produced the following rock types:

     (a)   sandstone with shell fragments
     (b)   shale with plant fossils
     (c)   sedimentary breccia
     (d)   greywacke
     (e)   oolitic limestone

10. What is an artesian basin? What limitations are placed upon using artesian water?

11. Define each of the following:

(a) porphyroblastic;
(b) augen;
(c) diagenesis;
(d) metamorphic aureole;
(e) facies;
(f) protolith.

12. What is a polymorph? Give an example and explain its significance in the identification of metamorphic rocks.

13. What is meant by the term retrograde metamorphism? How could you distinguish a retrograde metamorphic rock from one which had been formed progressively to that same grade?

14. Outline the metamorphic rocks which would be formed from a granite intrusion into dolomite. How could a petrologist distinguish between these and those formed by a similar intrusion into limestone?

15. Use the internet to research the:

(a) the economic use of (i) igneous rocks and
                          (ii) sedimentary rocks
(b) the metamorphic rocks in your immediate location and
(c) the economic use of metamorphic rocks or materials in them.

Answers to Multichoice Questions

Q1. B Q2. D Q3. C Q4. D Q5. C Q6. B Q7. D Q8. A Q9. A Q10. D

Reading List

Blatt, H., G. Middleton & R. Murray. (1980). *Origin of Sedimentary Rocks*. Saddle River, NJ: Prentice-Hall. ISBN: 0-13-642710-3.

Blatt, H. and Tracy, R. J. (1996) *Petrology*, W.H.Freeman, 2nd ed., p.355 ISBN 0-7167-2438-3.

Collinson, J.D. Thompson, D.B. (1988). *Sedimentary Structures* (2nd edit.). Boston: Unwin Hyman.

Fichter, L.S. (2000). *Sedimentary Rocks*. Department of Geology and Environmental Science James Madison University, Harrisonburg, Virginia. http://csmres.jmu.edu/geollab/fichter/SedRx/index.html

Frost, B.R. & Frost, C.D. (2013). *Essentials of Igneous and metamorphic Petrology*. Ambridge University Press. ISBN-10 1107696291

Geology.com. (2015). *Rocks: Igneous, Metamorphic and Sedimentary.*(Great database) http://geology.com/rocks/

Gill, R. (2007). *Igneous Rocks and Processes – A Practical Guide*. Chicester UK: Wiley-Blackwell. ISBN 9780632063772

Gillen, Cornelius. (1982). *Metamorphic Geology : an Introduction to Tectonic and Metamorphic Processes*, London; Boston: G. Allen & Unwin ISBN 978-0045510580.

Jahns, R.H. (2015). *Igneous Rock Geology*. Encyclopaedia Britannica.
http://www.britannica.com/science/igneous-rock

James, N.P & Jones, B. (2015). *Origin of Carbonate Sedimentary Rocks*. Chicester, UK: Wiley-Blackwell. 464 pp. ISBN-10: 1118652738

Jerram, D. 2001. *The Field Description of Igneous Rocks*. Chicester UK: Wiley-Blackwell. ISBN: 9780470022368

Le Maitre, R.W. (Edit.). (2005). *Igneous Rocks: A Classification and Glossary of Terms*. Cambridge University Press. 256 pages. ISBN-10: 0521619483

Michna, P. (2015). *Igneous Rocks*. Earthsci.org.
http://earthsci.org/education/teacher/basicgeol/igneous/igneous.html

Pettijohn, W. (2005). *Sedimentary Rocks*. New Delhi: CBS Publications. 220 pp, ISBN-10 812390875X

Pidwirny M. & Jones, S. (2014). *Characteristics of Igneous Rocks*. *Fundamentals eBook*. PhysicalGeography.net. Okanagan: University of British Columbia.
http://www.physicalgeography.net/fundamentals/10e.html

Prothero, D. R. & Schwab, F. (2004). *Sedimentary Geology : An Introduction to Sedimentary Rocks and Stratigraphy* (2nd ed.). New York: Freeman. ISBN 978-0-7167-3905-0.

Stow, D. A. V. (2005). *Sedimentary Rocks in the Field*. Burlington, MA: Academic Press. ISBN 978-1-874545-69-9.

Tucker, M.E. (2001). *Sedimentary Petrology: An Introduction to the Origin of Sedimentary Rocks*. Chicester, UK. Wiley-Blackwemm 272 pages. ISBN-10 1118652738

United States Geological Survey, (2014). *US Geology in the Parks: Metamorphic Rocks*. http://geomaps.wr.usgs.gov/parks/rxmin/rock3.html

Vernon, Ronald Holden. (2008). *Principles of Metamorphic Petrology*, Cambridge University Press ISBN 978-0521871785

Wicander R. & Munroe J. (2005). *Essentials of Geology*. Cengage Learning. *pp. 174-177*. ISBN 9780495013655.

Winter, J.D. (2009). *Principles of Igneous and Metamorphic Petrology*. New Jersey: Prentice Hall. 720 pages. ISBN-10 0321592573

Key Terms Index
(Page numbers in brackets)

accretion (160) is the joining of rock material as tectonic plates collide.

accessory minerals (45) are minerals in igneous rocks which formed at the same time as the primary minerals but occur in small amounts e.g. magnetite, chromite, apatite and sphene.

acidic igneous rocks (45) refers to an old chemical classification whereas the amount of silica (silicon dioxide) in the rock, like most chemical oxides would dissolve in water to form acids. Acid igneous rocks have high silica content.

adsorption (118) is the adherence of sub-surface fluids such as oil and liquefied gas to the walls of pore spaces.

aeolian (72) sediments have been deposited by the wind.

amygdules (49) are gas bubbles or vesicles filled with secondary minerals which crystalized later from remaining hot solutions.

angle of repose (124) is the angle that the foresets in current bedding make to the horizontal bedding plane and represents the steep side on the lee (away from the wind or water current) of a dune.

aphanitic (43) is a texture in igneous rocks where there are no or only a few small crystals generally visible to the naked eye, but under the microscope with polarised light, well-formed crystals are seen e.g. basalt.

aquicludes (112) are rocks which are impermeable and will not allow water to pass through. Some may be porous and hold moisture.

aquifers (112) sedimentary rock or alluvium which is permeable (will allow water to pass through) and will contain water.

aquitards (112) are aquifers with poor permeability so that water will only slowly pass through.

arenites (77) from the latin *arena* for sand, and is a sedimentary clastic rock with sand clast size between 0.0625 mm and 2 mm and contain less than 15% matrix.

artesian bore (113) is a drilled hole into an aquifer which allows the pressurized groundwater to rise up under its own pressure. It also often includes dissolved natural gas and minerals and is usually only suitable for stock.

artesian system (112) is a sequence of sedimentary rocks which includes water-bearing aquifers and constraining aquicludes.

augen (145) is German for eye and refer to concentrations of minerals in some gneisses which look like eyes.

aureole (147) is the zone around a hot igneous body that will cause contact metamorphism in the surrounding country rock. Its size depends upon the heat given out from the igneous both and can vary from a few millimetres to many kilometres.

bar (Table on p.134,139) is a non- SI unit of pressure and is approximately equal to 100 kilopascals of pressure or the atmospheric pressure 100 metres above sea-level.

basic igneous rocks (46) have little or no free silica but are rich in mafic minerals.

basic magma (33) is molten rock with very little silica (quartz) content.

batholiths (29) intrusive structure with area greater than 60 square kilometres.

bosses (29) are very small intrusive structures usually circular in shape and with vertical sides.

Bowen's Reaction Series (39) outlines the sequence of crystallisation of minerals within a magma as the temperature drops.

breccia (61) are rocks formed with broken clasts indicating that they have not been rolled around in the transporting medium e.g. as pyroclastics or from glaciers, landslides or faults.

burial metamorphism (146) is the change in rock or sediment due to the weight of material upon it. Usually it is an extended form of diagenesis (the compacting process of sediments).

cataclastic rocks (75) have been formed from the collapse of the ceiling within large caves.

cataclastic metamorphism (159) occurs as a result of mechanical deformation and heating by friction when two bodies of rock slide past one another along a fault or shear zone.

chemistry (19) refers to the chemical composition of the mineral i.e. what elements and chemical bonding make up the mineral.

christmas tree (116) is a valve system with multiple outlets which can be attached to an oil rig pipe at base to take the oil once it has come to the surface.

clasts (68) are the particles and fragments found in sedimentary rocks e.g. pebbles, sand and finer fragments.

clastic (74) sedimentary rocks are rocks having visible particles formed by their abrasion, transportation, deposition, compaction and cementation.

cleats (118) are the fractures within coal seams which often contain coal seam gases.

cleavage (9) is the way that some minerals split along natural, flat planes of weakness when gently struck.

coal seam gas (118) consists mainly of methane gas and a little carbon dioxide and is found within the cleats of coal beds. Once fractured and released it can be pumped to the surface and used as a fuel.

colour (5) is shown by the wavelength of light emerging from within the internal structure of the mineral.

conchoidal fracture (13) is when a mineral or rock breaks with a sharp edge which has concentric circles like those seen on the edges of broken shell hence the name *konche* for a mussel shell.

contact metamorphism (147) the change in rock by heat due to a nearby igneous body often with minimum pressure.

continuous reaction (39) occurs as magma cools and minerals crystallise. Ions, charged atoms of elements left in the residual solution, which has not yet crystallised, may move into the formed crystals and replace some of the ions within the crystal. This goes on continuously giving crystals with zones of different compositions e.g. plagioclase feldspars.

cratons (159) old shield areas of continents having ancient, tectonically stable crystalline rocks and usually forming the core of that continent.

cryptocrystalline (90) refers to rocks which have crystals or clasts which can only be seen under a petrological microscope.

crystal family (13) is a useful only when specimens have distinct, well-formed crystals, large enough to show distinct crystal faces and the angles between them.

current bedding (122) is shown within a sedimentary bed as a series of almost parallel lines which represent the successive layers of sediment push up the sides of dunes or ripples by a current of water or air.

diagenesis (147) is the physical and chemical changes occurring during the conversion of sediment to sedimentary rock.

diaphaneity (7) is the way that light passes through the specimen in normal thickness.

diapir (91) is an underground deposit of mineral which can be forced upwards by tectonic forces to form as a large dome e.g. salt domes.

discontinuous reaction (39) occurs with other minerals which form as solid crystals but then re-dissolve in the residual liquid of the magma, react and then recrystallize as new minerals at a lower temperature e.g. ferromagnesium minerals.

dykes (29) thin rectangular structure which intrude surrounding rock, country rock, by cutting across its layers.

extrusive (41) refers to a texture or igneous rock which shows small crystals due to cooling on the Earth's surface.

facies (135) is the assemblage of specific minerals which are formed at specific temperatures and/or pressures (each mineral is termed an index mineral).

felsic (27) are those minerals which are light in colour such as feldspars (**fel**-) and high in silicates (-**sic**) e.g. orthoclase, plagioclase, quartz.

flow lines (48) are streaks or partial lines seen within an extrusive igneous rock formed by different concentrations of material aligned when the molten rock was flowing.

fluvial (72) pertaining to rivers.

foliated (135) is a textural term denoting the semi-layers of flattened minerals, often as flakes found in many regional metamorphic rocks. Contact metamorphic rocks are usually non-foliated.

foresets (122) are the lines of sediment seen within current bedding.

fracture (12) is the way the mineral breaks up roughly, but not along smooth planes of weakness.

friable (99) a term used in all three rock types meaning that the rock is easily broken apart.

geotherm (137) is the curve of the geothermal gradient.

geothermal gradient (137) is the increase in heat per unit depth into the Earth and often defines the type of mineral changes which occur.

gneissic (144) is a coarsely foliated metamorphic rock texture with very large, coarse-grained flattened crystals of micas, feldspars and quartz.

granular (141) or granoblastic texture in metamorphic rocks is a non-foliated texture with well-formed crystals of equal size and well-defined edges.

habit (2) describes the overall appearance and arrangement of the crystals of the mineral specimen.

hardness (8) is the resistance to scratching when tested with a standard set of items (Mohs' Scale) having a relative strength.

hawaiite (36) is a type of basic lava common to some shield volcanoes such as those in Hawai'i

heft (18) is an approximate estimation of the heaviness of a mineral.

holohyaline (44) is an igneous rock texture, often called glassy, which contains no crystals e.g. obsidian and other volcanic glasses.

hornfelsic (141) is a texture in metamorphic rocks which is non-foliated texture, fine grained and with randomly oriented crystals uniform in size and shape.

hydrated (91) refers to water content within the lattice of some minerals e.g. gypsum $CaSO_4$. $2H_2O$ is the hydrated form of anhydrite $caso_4$.

hydrocarbons (115) are chemical compounds of hydrogen and oxygen and range from volatile gases such as methane (CH_4) to large chain and ring compounds found in heavy liquids such as oils and tars which contain many atoms.

hydrothermal metamorphism (158) is the hydrothermal alteration of country rocks by the intrusion of very hot solutions, often charged with carbon dioxide, derived from some nearby igneous body from which these solutions have been expelled.

idiomorphic crystals (140) are well-shaped crystals with regular faces within a metamorphic rock.

igneous rocks (1) are rocks formed from the cooling of magma below the surface or from lava on the surface.

index minerals (149) occur in metamorphic rock facies which appear at the specific temperatures for each type of parent rock.

intermediate igneous rocks (45) have only a middle range of silica, about 50%–60%.

intrusive (41) refers to an igneous rock texture of large, well-formed crystals cooled below the Earth's surface.

intrusive igneous (28) refers to rocks formed below the Earth's surface by the cooling of molten rock called magma.

isochemical metamorphism (140) from the Greek: *iso* - same and refers to those rocks which maintain their basic chemical composition even after metamorphism e.g. marble is still chemically the same as its original protolith limestone, both are mainly calcium carbonate. Most originally came from rocks composed of only one mineral, monomineralic.

kelly (116) is a polygonal slot into which the lengths or sticks of drilling pipe are attached

laccoliths (29) mushroom-shaped structures which squeeze between country rock layers and then push up as domes.

lacustrine (72) pertaining to lakes.

laminae (86) are the fine layers seen within some sedimentary rocks.

lava (1) is molten rock on the Earth's surface which cools to form extrusive igneous rock.

Law of Superposition (67) states that in a sequence of layered beds, the youngest is always deposited over the oldest.

lithification (68) is the complex process which forms rock. In sedimentary rocks, it involves deposition, compaction and cementation of particles.

lithostatic pressure (146) the weight of the rocks above causing a rock or sediment to undergo metamorphism.

lopoliths (30) intrude between layers but then collapse in the middle forming thin, basin-like structures.

lustre (6) is the way which light reflects off the surfaces of the mineral specimen.

lutites (77) an old term for fine grained sedimentary rocks such as claystones and siltstones.

mafic (27) or ferro-magnesium minerals are those minerals which contain large amounts of and **mag**nesium and iron (Latin : *ferrum*) within their structure. They are usually dark green or black in colour e.g. olivine and hornblende.

magma (1,26) is molten material from below the Earth's crust which may cool to form extrusive igneous rock.

magmatic differentiation (50) may occur as a magma cools deep below the surface and it undergoes various processes which cause different minerals to form and concentrate in different parts of the magma chamber, often as layers.

matrix (30,42) or groundmass is the background of smaller crystals in which the larger phenocrysts are embedded in porphyritic igneous rocks and in some metamorphic rocks. It can also refer to the background clasts within sedimentary rocks.

metamorphic rocks (2) named from the Greek *meta* for change and *morphe* for form and applies to all rocks changed by heat and/or pressure.

metasomatism (140) these are the changes made within rocks by the metamorphic process.

metastable (136) is a term referring to those minerals which remain within metamorphic rocks even though new temperatures and pressures may occur.

migmatites (137) are very high grade, high temperature metamorphic rocks which have undergone some re-melting and so contain some igneous rock features.

modal analysis (75) is the process of analysing sediments and their rocks by sieving and determining the percentage of the different sizes of clasts.

Mohs' scale (8) a list of hardness test minerals developed by Friedrich Mohs (German: 1773-1839) which have been given values of hardness and arranged in order from softest (1) to hardest (10).

monomineralic (89) are rocks which are comprised of mainly one mineral e.g. calcite in limestone.

non-clastic (74) sedimentary rocks have no visible particles and are formed from chemical precipitates or from biological sources.

non-foliated (135) are metamorphic rocks where there appears not to be any orientation of crystals which are more granular than flaky.

oil shales (88) are dark grey-black in colour and have pores filled with the hydrocarbon kerogen and can be used as a fuel source.

oil traps (115) are reservoirs within porous sedimentary rocks such as sandstone which contain gas, oil and water. These fluids are contained within the reservoir rock by containing impermeable rocks such as shales.

oligomictic (82) are conglomerates which are composed of clasts of mainly one type of mineral or rock.

oolites (96) from the Greek: *òoion* for egg, are round spheres showing concentric internal structure formed on gently-sloped calcareous beaches.

paludal (72) refers to swamp or marsh environments.

permeability (109) is the ability to allow fluids such as water, oil and gas to pass through.

petroleum (115) is a sub-surface hydrocarbon mixture of crude oils and natural gases.

phaneritic (42) an igneous rock texture in which all crystals are large, well-formed and are usually interlocked e.g. granite.

phenocrysts (30,42) are large crystals grown within a small-crystal mass (matrix). Found in porphyritic rocks.

phyllitic (143) is a foliated metamorphic rock texture with fine grains of mica just visible giving it a slight sheen.

polarized light (43) is used by mineralogists to identify minerals in thin-section (rock cut then ground on a microscope slide so that it is transparent) with a petrological microscope. This instrument has two sheets of polaroid film – one below the stage which holds the slide and the other in the tube of the microscope. When the stage of the microscope is rotated, the light which has been oriented in one direction (polarized) passes through the specimen and is again polarized by the second polaroid filter. This causes several different patterns of birefringence as changing colours which can then be used to identify the mineral or determine a rocks composition. In earlier times, the naturally-occurring mineral iceland spar (transparent calcite) was used instead of stretched plastic of polaroid.

polymictic (82) conglomerates have a great variety of clast types. The homogeneity, or lack of it, is determined by the sources of the clasts.

polymorphs (154) are minerals which have the same chemical composition but different crystal lattices (e.g. aluminium silicate – Al_2SiO_5 for the minerals kyanite, andalusite and sillimanite).

porosity (106) is the ability of a rock or sediment to hold water because of its pore spaces. A measure of a rock's porosity is found as a percentage of the pore volume to the total volume of the rock.

porphyritic (42) igneous rock texture which has large, well-formed crystals (phenocrysts) set into a background (groundmass) of finer crystals e.g. hornblende porphyry.

porphyroblasts (144) are concentrations of minerals within metamorphic rocks e.g. garnets within a garnet schist.

primary minerals (45) are the majority of minerals in an igneous rock which formed first as the molten rock cooled.

primary sedimentary structures (119) are shapes which have been formed before the sediments have become rock, often as part of sedimentation and lithification.

protoliths (132) are the original igneous, sedimentary and metamorphic rocks which have undergone metamorphism.

pyroclastic (27) refers to extrusive igneous material ejected as broken fragments, ash and bomb-like structures from volcanoes.

pyrolite (36) material formed at subduction zones when crustal rocks and ultra-basic mantle material below partially melts.

regional metamorphism (151) is the change in rock due to heat and/or pressure.

retrograde metamorphism (158) is a reversal of the usual gradation which takes place with increasing metamorphism (prograde metamorphism) involving the reconstitution of a rock to a lower grade of metamorphic stability under decreasing temperatures (and usually pressures) with the addition of volatiles (such as water and carbon dioxide).

rhombohedral system (17) the rhombohedral system is defined by its crystal lattice rather than outward appearance and is a subset of the trigonal lattice system. It has all sides equal including the c-axis and the angle between the c-axis and the others is not 90^0.

rhyolite (37,59) is an acid, extrusive igneous rock common to lavas formed from a granitic magma.

rock cycle (66) is the great recycling process whereby the rocks of the Earth are created, destroyed and their components recycled to form new rocks.

roundness (70) is the degree of smoothing due to abrasion of sedimentary particles.

rudites (77) are coarse grained sedimentary rocks such as conclomerates.

saccharoidal (135,142) is a textural term indicating that the rock looks grainy like coarse sugar.

sanidine facies (150) is rare metamorphic facies containing the index mineral sanidine. It is a very high temperature facies, beyond the usual range of the highest temperatures normally encountered in contact metamorphism and is at the edge of complete re-melting of the rock to form magma.

schiller iridescence (62) from the German word schiller meaning to twinkle is an optical effect caused by the breaking up of white light into its several colours due to fine layers within a mineral, especially some feldspars.

schistose (143) is a foliated metamorphic rock texture with large, flaky crystals of micas, chlorites and sericite minerals giving a great variety of schists named after the main mineral e.g. biotite schist.

secondary minerals (45) are minerals in rocks which have been formed later by chemical weathering or alteration of the existing minerals e.g. clays, chlorite, calcite, haematite and serpentine.

secondary sedimentary structures (119) formed after or near the end of lithification.

sedimentary rocks (1) are rocks formed from the deposition, compaction and cementation of loose particles or organic remains.

sedimentary structures (118) internal features which may be seen within a layer or bed of sedimentary rock as opposed to some beds which are massive and appear to have no internal structure.

shield volcano (34) very large eruptive structures of rounded profile built up by successive layers of lava. Commonly formed within a tectonic plate by a hot mantle plume below.

shock metamorphism (159) or impact metamorphism describes the changes produced in country rock when an extra-terrestrial body, such as a meteorite or comet impacts with the Earth or if there is a very large volcanic explosion and ultra high pressures (and temperatures) are produced.

sills (29) thin rectangular structures which intrude country rock by pushing between layers.

slaty (142) is a foliated texture in metamorphic rocks consisting of very fine grains, often too small to see with the naked eye, with orientation of crystals giving flat layers which are easy to split apart.

sorting (71) the similarity in grain size of a rock due to the transportation of the sediment such that well-sorted rocks have particles (clasts) about the same size and shape.

specific gravity S.G. (18) is the density of the mineral compared to that of water which is 1.0 gram per cubic centimetre.

specific properties (19) refer to any unique features of the mineral which may be useful in quick identification e.g. magnetite can be attracted to a magnet.

sphericity (70) is the degree of shape of clast to approach that of a perfect sphere.

stick (116) are the lengths of drilling pipe which can be screwed together in a drilling rig.

stocks (29) are intrusive structures usually less than 60 square kilometres and circular in shape.

strata (67) are distinct layers within an individual sedimentary rock sequence.

stratovolcano (35) also called a composite volcano is formed by successive layers of ash and lava. Typically has the well-known cone shape with even slopes. Commonly formed at subduction zones where one plate is pushed below another, re-melting crust and surface sediments.

streak (6) is the colour of the powdered mineral made by firmly pressing the mineral across a white tile called a streak plate.

sub-artesian (114) underground water lacks hydrostatic pressure so must be pumped to the surface.

sutured (141) the edges of crystals have crinkled edges looking as though they have been stitched.

tectonic (35,138) refers to the gigantic forces within the Earth.

tectonic plates (160) are the mosaic pieces which form the surface of the Earth and are in constant interaction and movement.

tephra (27) all pyroclastic particles, irrespective of size and shape.

terrestrial (72) in geology this refers to sedimentary environments which are non-marine and found on land.

trigonal crystal system (17) is defined by its lattice, not appearance and has all axes equal and all angles equal (but not 90^0).

triple point (154) is a place on a graph of temperatures and pressure (phase diagram) where three phases (or in this case index minerals of the same composition) can co-exist.

turbidity current (83) is an underwater avalanche coming of the Continental Shelf and flowing down the Continental Slope. It consists of rock debris (regolith) and water.

ultra-basic (36) rock material is very rich in iron-magnesium minerals but has little or no quartz.

ultra-basic igneous rocks (46) are those rocks formed largely of ultra-basic minerals by igneous action.

varved shales (88) from the Swedish: *varv* for layers, are produced when very fine glacial till is washed out of a glacier and into lakes which may form in the valley beyond their front or snout.

vesicles (48) are round to elongated gas bubbles found in volcanic rocks due to the rising of gas bubbles from within the molten lava.

ventifacts (69) are angular rocks or rock formations shaped by the wind.

viscosity (111) is the frictional drag of a fluid – it "stickiness".

Wentworth - Udden scale (76) is a graded scale of particle size of clasts within sedimentary rocks and is used in their analysis and classification.

xenoblastic crystals (140) with no regular crystal faces within a metamorphic rock.

This book is also available in electronic format which can be purchased at amazon.com for Kindle or other electronic devices such as PCs and iPad using the free Kindle App. Books in the series **ADVENTURES IN EARTH SCIENCE** are available from Felix Publishing, Australia (info@felixpublishing.com) and include:

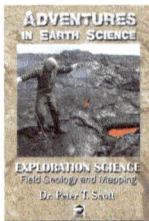

EXPLORATION SCIENCE
Field Geology &
Mapping

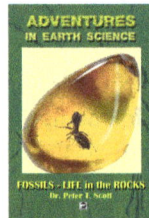

FOSSILS – LIFE in the ROCKS

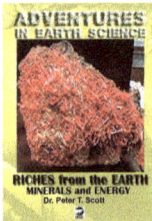

RICHES from the EARTH
Minerals & Energy

A DANGEROUS PLANET
Volcanoes &
Earthquakes

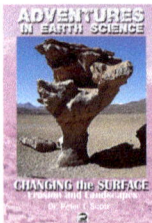

ROCKS – BUILDING the EARTH

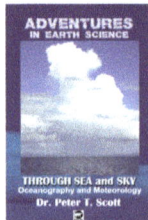

THROUGH SEA and SKY
Oceanography &
Meteorology

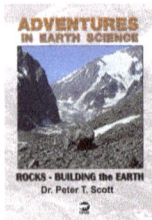

CHANGING the SURFACE
Erosion & Landscapes

BEYOND PLANET EARTH
An Introduction to Astronomy